FORD CONSUL · ZEPHYR ZODIAC Mk.1 & 11 1950 - 1962

Compiled by
R.M. Clarke

ISBN 1 85520 0589

Distributed by
Brooklands Book Distribution Ltd.
'Holmerise', Seven Hills Road,
Cobham, Surrey, England

Printed in Hong Kong

BROOKLANDS BOOKS

BROOKLANDS ROAD TEST SERIES

AC Ace & Aceca 1953-1983
Alfa Romeo Alfasud 1972-1984
Alfa Romeo Alfetta Coupes GT. GTV. GTV6 1974-1987
Alfa Romeo Giulia Berlinas 1962-1976
Alfa Romeo Giulia Coupes 1963-1976
Alfa Romeo Spider 1966-1990
Allard Gold Portfolio 1937-1958
Alvis Gold Portfolio 1919-1969
American Motors Muscle Cars 1966-1970
Aston Martin Gold Portfolio 1972-1985
Austin Seven 1922-1982
Austin A30 & A35 1951-1962
Austin Healey 100 & 100/6 Gold Portfolio 1952-1959
Austin Healey 3000 Gold Portfolio 1959-1967
Austin Healey 'Frogeye' Sprite Col No.1 1958-1961
Austin Healey Sprite 1958-1971
Avanti 1962-1983
BMW Six Cylinder Coupes 1969-1975
BMW 1600 Col. 1 1966-1981
BMW 2002 1968-1976
Bristol Cars Gold Portfolio 1946-1985
Buick Automobiles 1947-1960
Buick Muscle Cars 1965-1970
Buick Riviera 1963-1978
Cadillac Automobiles 1949-1959
Cadillac Automobiles 1960-1969
Cadillac Eldorado 1967-1978
High Performance Capris Gold Portfolio 1969-1987
Chevrolet Camaro SS & Z28 1966-1973
Chevrolet Camaro & Z-28 1973-1981
High Performance Camaros 1982-1988
Camaro Muscle Cars 1966-1972
Chevrolet 1955-1957
Chevrolet Corvair 1959-1969
Chevrolet Impala & SS 1958-1971
Chevrolet Muscle Cars 1966-1971
Chevelle and SS 1964-1972
Chevy Blazer 1969-1981
Chevy EL Camino & SS 1959-1987
Chevy II Nova & SS 1962-1973
Chrysler 300 1955-1970
Citroen Traction Avant Gold Portfolio 1934-1957
Citroen DS & ID 1955-1975
Citroen SM 1970-1975
Citroen 2CV 1949-1988
Shelby Cobra Gold Portfolio 1962-1969
Cobras & Replicas 1962-1983
Chevrolet Corvette Gold Portfolio 1953 1962
Corvette Stingray Gold Portfolio 1963-1967
High Performance Corvettes 1983-1989
Daimler SP250 Sport & V-8250 Saloon Gold Portfolio 1959-1969
Datsun 240Z 1970-1973
Datsun 280Z & ZX 1975-1983
De Tomaso Collection No.1 1962-1981
Dodge Charger 1966-1974
Dodge Muscle Cars 1967-1970
Excalibur Collection No.1 1952-1981
Facel Vega 1954-1964
Ferrari Cars 1946-1956
Ferrari Cars 1973-1977
Ferrari Dino 1965-1974
Ferrari Dino 308 1974-1979
Ferrari 308 & Mondial 1980-1984
Ferrari Collection No.1 1960-1970
Fiat-Bertone X1/9 1973-1988
Fiat Pininfarina 124 + 2000 Spider 1968-1985
Ford Automobiles 1949-1959
Ford Bronco 1966-1977
Ford Bronco 1978-1988
Ford Consul. Zephyr Zodiac MkI & II 1950-1962
Ford Cortina 1600E & GT 1967-1970
Ford Fairlane 1955-1970
Ford Falcon 1960-1970
Ford GT40 Gold Portfolio 1964-1987
Ford RS Escorts 1968-1980
High Performance Escorts Mk1 1968-1974
High Performance Escorts Mk II 1975-1980
High Performance Mustangs 1982-1988
Honda CRX 1983-1987
Hudson & Railton 1936-1940
Jaguar Cars 1957-1961
Jaguar Cars 1961-1964
Jaguar Mk2 1959-1969
Jaguar E-Type Gold Portfolio 1961-1971
Jaguar E-Type 1966-1971
Jaguar E-Type V-12 1971-1975
Jaguar XKE Collection No.1 1961-1974
Jaguar XJ6 1968-1972
Jaguar XJ6 Series II 1973-1979
Jaguar XJ6 & XJ12 Series III 1979-1985
Jaguar XJ12 1972-1980
Jaguar XJS Gold Portfolio 1975-1988
Jaguar XK120.XK140.XK150 Gold Portfolio 1948-1960
Jeep CJ5 & CJ6 1960-1976
Jeep CJ5 & CJ7 1976-1986
Jensen Cars 1946-1967
Jensen Cars 1967-1979
Jensen Interceptor Gold Portfolio 1966-1986
Jensen Healey 1972-1976
Lamborghini Cars 1964-1970
Lamborghini Cars 1970-1975
Lamborghini Countach Col No.1 1971-1982
Lamborghini Countach & Urraco 1974-1980
Lamborghini Countach & Jalpa 1980-1985
Lancia Stratos 1972-1985
Land Rover 1948-1973 - A Collection
Land Rover Series II & IIa 1958-1971
Land Rover Series III 1971-1985
Land Rover 90 & 110 1983-1989
Lincoln Gold Portfolio 1949-1960
Lincoln Continental 1961-1969
Lotus and Caterham Seven Gold Portfolio 1957-1989
Lotus Cortina Gold Portfolio 1963-1970
Lotus Elan Gold Portfolio 1962-1974
Lotus Elan Collection No.2 1963-1972
Lotus Elite 1957-1964
Lotus Elite & Eclat 1974-1982
Lotus Turbo Esprit 1980-1986
Lotus Europa 1966-1975
Lotus Europa Collection No.1 1966-1974
Lotus Seven Collection No.1 1957-1982
Marcos Cars 1960-1988
Maserati 1965-1970
Maserati 1970-1975
Mazda RX-7 Collection No.1 1978-1981
Mercedes 190 & 300SL 1954-1963
Mercedes 230/250/280SL 1963-1971
Mercedes Benz SLs & SLCs Gold Portfolio 1971-1989
Mercedes Bens Cars 1949-1954
Mercedes Bens Cars 1954-1957
Mercedes Bens Cars 1957-1961
Mercedes Bens Competion Cars 1950-1957
Mercury Muscle Cars 1966-1971
Metropolitan 1954-1962
MG TC 1945-1949
MG TD 1949-1953
MG TF 1953-1955
MG Cars 1959-1962
MGA Roadsters 1955-1962
MGA Collection No.1 1955-1982
MGB Roadsters 1962-1980
MGB GT 1965-1980
MG Midget 1961-1980
Mini Cooper Gold Portfolio 1961-1971
Mini Moke 1964-1989
Mini Muscle Cars 1961-1979
Mopar Muscle Cars 1964-1967
Mopar Muscle Cars 1968-1971
Morgan Three-Wheeler Gold Portfolio 1910-1952
Morgan Cars 1960-1970
Morgan Cars Gold Portfolio 1968-1989
Morris Minor Collection No.1
Mustang Muscle Cars 1967-1971
Oldsmobile Automobiles 1955-1963
Old's Cutlass & 4-4-2 1964-1972
Oldsmobile Muscle Cars 1964-1971
Oldsmobile Toronado 1966-1978
Opel GT 1968-1973
Packard Gold Portfolio 1946-1958
Pantera Gold Portfolio 1970-1989
Plymouth Barracuda 1964-1974
Plymouth Muscle Cars 1966-1971
Pontiac Tempest & GTO 1961-1965
Pontiac GTO 1964-1970
Pontiac Firebird 1967-1973
Pontiac Firebird and Trans-Am 1973-1981
High Performance Firebirds 1982-1988
Pontiac Fiero 1984-1988
Pontiac Muscle Cars 1966-1972
Porsche 356 1952-1965
Porsche Cars in the 60's
Porsche Cars 1960-1964
Porsche Cars 1964-1968
Porsche Cars 1968-1972
Porsche Cars 1972-1975
Porsche Turbo Collection No.1 1975-1980
Porsche 911 1965-1969
Porsche 911 1970-1972
Porsche 911 1973-1977
Porsche 911 Carrera 1973-1977
Porsche 911 Turbo 1975-1984
Porsche 911 SC 1978-1983
Porsche 914 Gold Portfolio 1969-1976
Porsche 914 Collection No.1 1969-1983
Porsche 924 Gold Portfolio 1975-1988
Porsche 928 1977-1989
Porsche 944 1981-1985
Range Rover Gold Portfolio 1970-1988
Reliant Scimitar 1964-1986
Riley 11/2 & 21/2 Litre Gold Portfolio 1945-1955
Rolls Royce Silver Cloud 1955-1965
Rolls Royce Silver Shadow 1965-1981
Rover P4 1949-1959
Rover P4 1955-1964
Rover 3 & 3.5 Litre 1958-1973
Rover 2000 + 2200 1963-1977
Rover 3500 1968-1977
Rover 3500 & Vitesse 1976-1986
Saab Sonett Collection No.1 1966-1974
Saab Turbo 1976-1983
Shelby Mustang Muscle Cars 1965-1970
Stubebaker Gold Portfolio 1947-1966
Stubebaker Hawks & Larks 1956-1963
Sunbeam Tiger & Alpine Gold Portfolio 1959-1967
Thunderbird 1955-1957
Thunderbird 1958-1963
Thunderbird 1964-1976
Toyota Land Cruiser 1956-1984
Toyota MR2 1984-1988
Triumph 2000. 2.5. 2500 1963-1977
Triumph GT6 1966-1974
Triumph Spitfire 1962-1980
Triumph Spitfire Col No.1 1962-1982
Triumph Stag 1970-1980
Triumph Stag Collection No.1 1970-1984
Triumph TR2 & TR3 1952-60
Triumph TR4-TR5-TR250 1961-1968
Triumph TR6 1969-1976
Triumph TR6 Collection No.1 1969-1983
Triumph TR7 & TR8 1975-1982
Triumph Herald 1959-1971
Triumph Vitesse 1962-1971
TVR Gold Portfolio 1959-1988
Volkswagen Cars 1936-1956
VW Beetle Collection No.1 1970-1982
VW Golf GTi 1976-1986
VW Karmann Ghia 1955-1982
VW Kubelwagen 1940-1975
VW Scirocco 1974-1981
VW Bus. Camper. Van 1954-1967
VW Bus. Camper. Van 1968-1979
VW Bus. Camper. Van 1979-1989
Volvo 120 1956-1970
Volvo 1800 1960-1973

BROOKLANDS ROAD & TRACK SERIES

Road & Track on Alfa Romeo 1949-1963
Road & Track on Alfa Romeo 1964-1970
Road & Track on Alfa Romeo 1971-1976
Road & Track on Alfa Romeo 1977-1989
Road & Track on Aston Martin 1962-1984
Road & Track on Auburn Cord and Duesenburg 1952-1984
Road & Track on Audi & Auto Union 1952-1980
Road & Track on Audi 1980-1986
Road & Track on Austin Healey 1953-1970
Road & Track on BMW Cars 1966-1974
Road & Track on BMW Cars 1975-1978
Road & Track on BMW Cars 1979-1983
Road & Track on Cobra, Shelby & GT40 1962-1983
Road & Track on Corvette 1953-1967
Road & Track on Corvette 1968-1982
Road & Track on Corvette 1982-1986
Road & Track on Datsun Z 1970-1983
Road & Track on Ferrari 1950-1968
Road & Track on Ferrari 1968-1974
Road & Track on Ferrari 1975-1981
Road & Track on Ferrari 1981-1984
Road & Track on Fiat Sports Cars 1968-1987
Road & Track on Jaguar 1950-1960
Road & Track on Jaguar 1961-1968
Road & Track on Jaguar 1968-1974
Road & Track on Jaguar 1974-1982
Road & Track on Jaguar 1983-1989
Road & Track on Lamborghini 1964-1985
Road & Track on Lotus 1972-1981
Road & Track on Maserati 1952-1974
Road & Track on Maserati 1975-1983
Road & Track on Mazda RX7 1978-1986
Road & Track on Mercedes 1952-1962
Road & Track on Mercedes 1963-1970
Road & Track on Mercedes 1971-1979
Road & Track on Mercedes 1980-1987
Road & Track on MG Sports Cars 1949-1961
Road & Track on MG Sprots Cars 1962-1980
Road & Track on Mustang 1964-1977
Road & Track on Nissan 300-ZX & Turbo 1984-1989
Road & Track on Peugeot 1955-1986
Road & Track on Pontiac 1960-1983
Road & Track on Porsche 1961-1967
Road & Track on Porsche 1968-1971
Road & Track on Porsche 1972-1975
Road & Track on Porsche 1975-1978
Road & Track on Porsche 1979-1982
Road & Track on Porsche 1982-1985
Road & Track on Porsche 1985-1988
Road & Track on Rolls Royce & B'ley 1950-1965
Road & Track on Rolls Royce & B'ley 1966-1984
Road & Track on Saab 1955-1985
Road & Track on Toyota Sports & GT Cars 1966-1984
Road & Track on Triumph Sports Cars 1953-1967
Road & Track on Triumph Sports Cars 1967-1974
Road & Track on Triumph Sports Cars 1974-1982
Road & Track on Volkswagen 1951-1968
Road & Track on Volkswagen 1968-1978
Road & Track on Volkswagen 1978-1985
Road & Track on Volvo 1957-1974
Road & Track on Volvo 1975-1985
Road & Track - Henry Manney at Large and Abroad

BROOKLANDS CAR AND DRIVER SERIES

Car and Driver on BMW 1955-1977
Car and Driver on BMW 1977-1985
Car and Driver on Cobra, Shelby & Ford GT 40 1963-1984
Car and Driver on Corvette 1956-1967
Car and Driver on Corvette 1968-1977
Car and Driver on Corvette 1978-1982
Car and Driver on Corvette 1983-1988
Car and Driver on Datsun Z 1600 & 2000 1966-1984
Car and Driver on Ferrari 1955-1962
Car and Driver on Ferrari 1963-1975
Car and Driver on Ferrari 1976-1983
Car and Driver on Mopar 1956-1967
Car and Driver on Mopar 1968-1975
Car and Driver on Mustang 1964-1972
Car and Driver on Pontiac 1961-1975
Car and Driver on Porsche 1955-1962
Car and Driver on Porsche 1963-1970
Car and Driver on Porsche 1970-1976
Car and Driver on Porsche 1977-1981
Car and Driver on Porsche 1982-1986
Car and Driver on Saab 1956-1985
Car and Driver on Volvo 1955-1986

BROOKLANDS PRACTICAL CLASSICS SERIES

PC on Austin A40 Restoration
PC on Land Rover Restoration
PC on Metalworking in Restoration
PC on Midget/Sprite Restoration
PC on Mini Cooper Restoration
PC on MGB Restoration
PC on Morris Minor Restoration
PC on Sunbeam Rapier Restoration
PC on Triumph Herald/Vitesse
PC on Triumph Spitfire Restoration
PC on VW Beetle Restoration
PC on 1930s Car Restoration

BROOKLANDS MOTOR & THOROGHBRED & CLASSIC CAR SERIES

Motor & T & CC on Ferrari 1966-1976
Motor & T & CC on Ferrari 1976-1984
Motor & T & CC on Lotus 1979-1983

BROOKLANDS MILITARY VEHICLES SERIES

Allied Mil. Vehicles No.1 1942-1945
Allied Mil. Vehicles No.2 1941-1946
Dodge Mil. Vehicles Col. 1 1940-1945
Military Jeeps 1941-1945
Off Road Jeeps 1944-1971
Hail to the Jeep
US Military Vehicles 1941-1945
US Army Military Vehicles WW2-TM9-2800

BROOKLANDS HOT ROD RESTORATION SERIES

Auto Restoration Tips & Techniques
Basic Bodywork Tips & Techniques
Basic Painting Tips & Techniques
Camaro Restoration Tips & Techniques
Custom Painting Tips & Techniques
Engine Swapping Tips & Techniques
How to Build a Street Rod
Mustang Restoration Tips & Techniques
Performance Tuning - Chevrolets of the '60s
Performance Tuning - Ford of the '60s
Performance Tuning - Mopars of the '60s
Performance Tuning - Pontiacs of the '60s

CONTENTS

ACKNOWLEDGEMENTS

In recognition of today's interest in the popular British family saloon models of the 1950s, this new Brooklands Books' title looks at the big-selling Ford Consul Zephyr Zodiac Mk1 and Mk2 range produced at Dagenham from 1950 to 1962.

We have asked Michael Allen, a leading authority on many of Ford of Britain's products, to write a few paragraphs as an introduction. Michael's own definitive work on the Dagenham large-car series, titled "Consul Zephyr Zodiac Executive: Fords Mk1 to 4", published by Motor Racing Publications, is also thoroughly recommended as further reading for all Ford enthusiasts.

We would not of course have been able to produce this book at all without the kind co-operation of the copyright holders of the magazine features reproduced within, and therefore our gratitude is extended to the proprietors of Autocar, Car Life, Cars, Car – South Africa, Classic & Sportscar, Modern Motor, Motor, Motor Life, Motor Sport, Practical Classics, Road & Track, Sports Cars Illustrated, and Wheels.

R.M. Clarke

With their MacPherson Strut front suspension, OHV oversquare engines, pendant foot pedals, and integral body/chassis construction, the new Consul (4-cylinder) and Zephyr Six which appeared on the Ford stand at the Motor Show in October 1950 represented much completely new thinking by Ford, and which, in part, was new to the industry as a whole. And yet for all this, these smooth newcomers remained overall very typically Ford in that they were beautifully simple, rugged, and of course were being offered at unbeatable prices in their segment of the market place.

A more luxurious Zephyr Zodiac saloon was added for 1954, its timely introduction allowing Ford to move a little further upmarket with their popular six-cylinder car as postwar home market restrictions were eased, and Britain's economy expanded. Convertible versions of the Consul and Zephyr Six had appeared by this time, and the range was spread further still late in 1954 when estate car conversions were offered by specialist coachbuilders E.D. Abbott. Annual increases in production were never quite sufficient to satisfy the demand for these Fords; a demand periodically stimulated by sporting successes, such as the outright victoyr gained on the Dutch Tulip Rally by a Consul in 1952, and the Zephyr Six outright win of the Monte Carlo Rally the following year.

Considerably enlarged Mk2 versions of the consul Zephyr and Zodiac took over early in 1956, with their rather more crisply styled bodywork nevertheless being built on similar principles to those of the Mk1 cars. The mechanical elements also remained substantially unchanged, but an increase in the engine capacities and raised overall gearing resulted in greater yet more effortless all round performance.

Popular to a greater extent than any other British large-car range of the time, the big Mk2 Fords took the Consul Zephyr Zodiac series past the 1,000,000 total production figure during 1961. A large slice of that million had sold overseas at a time when exports had never been more vital to the British economy, and amongst the pages of reprints which appear in this book perhaps the most interesting for many readers will be those taken from contemporary foreign magazines.

Michael Allen

Leeds
Yorkshire

Completely New Medium-sized 4- and 6-cylinder Cars of Extremely Original Design Announced from the Dagenham Factory

NEW LOOKS.—1951 additions to the Ford range are the 4-cylinder Consul and, nearer the camera, the Zephyr Six.

1951 CARS

Two New Ford Models for 1951

ONE of the rarest and most important of motoring events, the announcement of a new Ford range is an event which alone would ensure the success of the 1950 Motor Show. Those who have waited will not be disappointed by either the merit or the technical originality of the two new additions to the range of Dagenham-built Ford cars.

Taking their place alongside the Anglia and Prefect models which are renowned as providers of the most inexpensive and reliable form of long-distance personal transport, the new Consul and Zephyr Six offer greater comfort and carrying capacity, together with alternative degrees of enhanced performance. Similar in internal body dimensions, and sharing many technical features, the new models differ substantially in respect of appearances, engine size, and overall dimensions.

Both models use four-door bodywork of integral steel construction, supported at the rear on a ¾-floating axle and semi-elliptic leaf springs, at the front by a uniquely neat layout of flexible coil-spring I.F.S. Both models also employ power units of new pushrod o.h.v. design linked to 3-speed synchro-mesh gearboxes, the Consul engine being a four-cylinder rated at 16 h.p. but developing 47 b.h.p., the Zephyr Six of 50% greater displacement developing 68 b.h.p.

Modern full-width body styling, with emphasis on wind-screen and rear window width, gives maximum passenger and luggage capacity, electrical equipment is 12 volt pattern, brakes are hydraulic, and wheels are of 13-inch diameter.

NEW in appearance, the 1951 Fords make technical innovations, not merely for the sake of novelty but as a means to combine typically Ford low-cost motoring with new luxury and performance. Past practice, both of the Ford Motor Company and of other manufacturers, has obviously been studied closely, but wherever necessary both have been abandoned in favour of completely fresh ideas.

FORD DATA

Model	Consul	Zephyr Six
Engine Dimensions:		
Cylinders	4	6
Bore	79.37 mm.	79.37 mm.
Stroke	76.2 mm.	76.2 mm.
Cubic capacity	1,508 c.c.	2,262 c.c.
Piston area	30.7 sq. ins.	46.0 sq. ins.
Valves	Push-rod o.h.v.	Push-rod o.h.v.
Compression ratio ...	6.8 to 1	6.8 to 1
Engine Performance:		
Max. power	47 b.h.p.	68 b.h.p.
at	4,400 r.p.m.	4,000 r.p.m.
Max. b.m.e.p.	121 lb./sq. in.	122 lb./sq. in.
at	2,400 r.p.m.	2,000 r.p.m.
B.H.P. per sq. in. piston area	1.53	1.48
Peak piston speed ...	2,200 ft./min.	2,000 ft./min.
Engine Details:		
Carburetter	30 mm. downdraught	30 mm. downdraught
Ignition	12-volt coil	12-volt coil
Plugs: make and type ...	Champion NA8	Champion NA8
Fuel pump	A.C. Mechanical	A.C. Mechanical
Fuel capacity	9 gallons	9 gallons
Oil filter	A.C. full-flow	A.C. full-flow
Oil capacity	6½ pints	8 pints
Cooling system	Pump, fan and thermostat	Pump, fan and thermostat
Water capacity	16½ pints	22 pints
Electrical system ...	12-volt Lucas	12-volt Lucas
Battery capacity ...	45 amp/hr.	45 amp/hr.
Transmission:		
Clutch	8-in. single dry plate, with hydraulic control	8-in. single dry plate, with hydraulic control
Gear ratios: Top ...	4.625	4.375
2nd ...	7.598	7.187
1st	13.145	16.875
Rev. ...	17.839	16.875
Prop. shaft	Single open	Single open
Final drive	8/37 Hypoid Bevel	8/35 Hypoid Bevel
Chassis Details:		
Brakes	Girling hydraulic (2 L.S. front)	Girling hydraulic (2 L.S. front)
Brake drum diameter ...	9 in.	9 in.
Friction lining area ...	121 sq. ins.	121 sq. irs.
Suspension: Front ...	Coil spring I F.S.	Coil spring I.F.S.
Rear ...	Semi-elliptic leaf	Semi-elliptic leaf
Shock absorbers	Front: Direct-acting hydraulic Rear: Double-acting hydraulic	Front: Direct-acting hydraulic Rear: Double-acting hydraulic
Wheel type	Steel disc	Steel disc
Tyre size	5.90 x 13	6.40 x 13
Steering gear	Burman worm and peg	Burman worm and peg
Steering wheel	17 in. 3 spring spoke	17 in. 3 spring spoke
Dimensions:		
Wheelbase	8 ft. 4 ins.	8 ft. 8 ins.
Track: Front	4 ft. 2 ins.	4 ft. 2 ins.
Rear	4 ft. 1 in.	4 ft. 1 in.
Overall length	13 ft. 8¾ ins.	14 ft. 3¾ ins.
Overall width	5 ft. 4 ins.	5 ft. 4 ins.
Overall height	5 ft. 0¾ in.	5 ft. 0¾ in.
Ground clearance ...	6¾ ins.	6½ ins.
Turning circle	Right: 41 ft. Left: 39 ft. 6 ins.	Right: 43 ft. 4 ins. Left: 41 ft. 1 in.
Kerb weight	21¾ cwt.	23¼ cwt.

Model	Consul	Zephyr Six
Performance Data:		
Piston area, sq. ins. per ton	28.2	39.8
Brake lining area, sq. ins. per ton	111	105
Top gear m.p.h. per 1,000 r.p.m.	14.9	16.3
Top gear m.p.h. at 2,500 ft./min. piston speed ...	74.5	81.5
Litres per ton-mile, at kerb weight	2,790	3,580

SPEED.—High performance is promised by the new 2¼-litre Zephyr Six..

Two new cars are offered, a 1½-litre "four" and a 2¼-litre "six," both powered by modern short-stroke engines which promise to be economical and durable as well as lively. To venture estimates of probable performance, the Zephyr Six can be expected to exceed 80 m.p.h. by a considerable margin, to have at least equal top-gear acceleration to the 30 h.p. V8 Pilot, yet to give 25 m.p.g. or better fuel consumption. The Consul, on the other hand, is intended for those who want to be able to count on 30 m.p.g. fuel consumption, but should also be capable of excellent top-gear acceleration and of speeds in excess of 70 m.p.h.

Both cars are based upon fundamentally similar four-door saloon bodies of welded-up integral steel construction, no separate chassis frame being used in either case. The base structure of floor and body side rails is reinforced by U-section pressings above the rear axle, and has box-section forward extensions which vary in length as between 4- and 6-cylinder cars: in either instance, large side pressings forming the door frames, and another combining the windscreen frame and roof panel, provide reinforcement which gives complete rigidity to the centre-section of the car, whilst fixed panels alongside the o.h.v. engines and a bulkhead encircling the radiator extend this reinforcement to the extreme front of the car. Without excessive weight, expensive complication, or obstructed access to vital mechanical parts, immense structural strength and rigidity have been attained.

A New Front Suspension

For use in conjunction with this structure, a unique independent front wheel springing system has been evolved which, breaking right away from orthodox practice, is in fact notably straight-forward. First and foremost, it is a system which takes the principal loads directly, the weight of the car's front end being supported via rubber-insulated duplex conical steering swivel bearings on a flexible coil spring at each side of the car, these coil springs encircling telescopic guides on each of which a front wheel is mounted. Thus, any rise and fall of either wheel as it follows a rough road surface is absorbed directly by the coil spring: the telescopic guide, whose primary purpose is to maintain the wheel upright, is also a double-acting hydraulic shock absorber of very generous dimensions, controlling the motion of the flexible spring.

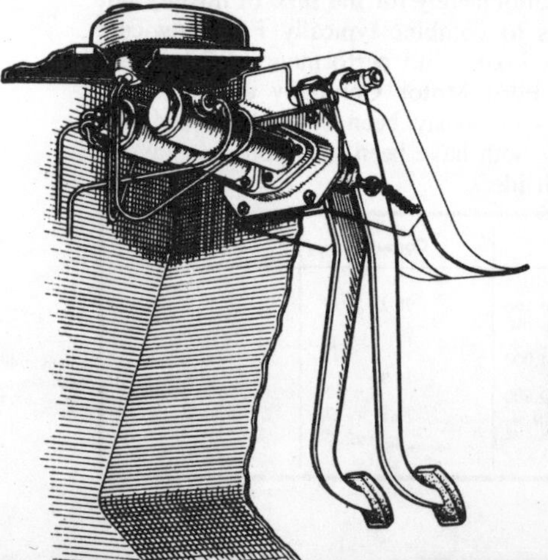

PAIRED PEDALS.—Top-mounted brake and clutch pedals are mounted directly adjacent to hydraulic master cylinders on the bulkhead.

Just below stub-axle level, the base of the telescopic sliding member is held by a ball-and-socket joint which takes the main drag and lateral forces involved in braking and cornering. This ball-joint rises and falls with the wheel, but is located laterally and fore-and-aft by what is, in effect, a transverse wishbone member swept back at an angle of approximately 25 degrees. One arm of this wishbone is an I-section transverse forging, rubber-bushed at its pivot on to a tubular frame cross member; the other is a backswept arm of circular section, actually the end of a rubber-mounted torsion anti-roll bar.

The layout of the complete assembly may be observed from the perspective drawings on these pages, which will be understood clearly when it is realised that spring deflection causes the anti-roll stabilizer to bend as well as to twist. Permitting slight track variations which large-section tyres should absorb readily, this novel suspension layout gives good roll resistance, is light and mechanically simple in relation to its strength and range of movement, and should transmit the minimum of road noise or shock.

Steering layout for use in conjunction with the Ford I.F.S. system is based upon use of a worm-and-peg steering gear of 14 : 1 ratio mounted to give transverse movement of the drop-arm, which latter is linked to a parallel-moving idler arm on the opposite side of the car. From points on the link between steering and idler arms, track-rod halves of a length chosen to suit the suspension geometry transmit steering movements independently to the two front wheels.

Correct matching of front and rear suspension designs is vital to the attainment of good riding and controllability, and on their new models the Ford Motor Company have adopted semi-elliptic leaf rear springs, of slightly higher natural frequency than the front springs, and provided with inserts between the leaf tips: control is provided by double-acting hydraulic shock absorbers.

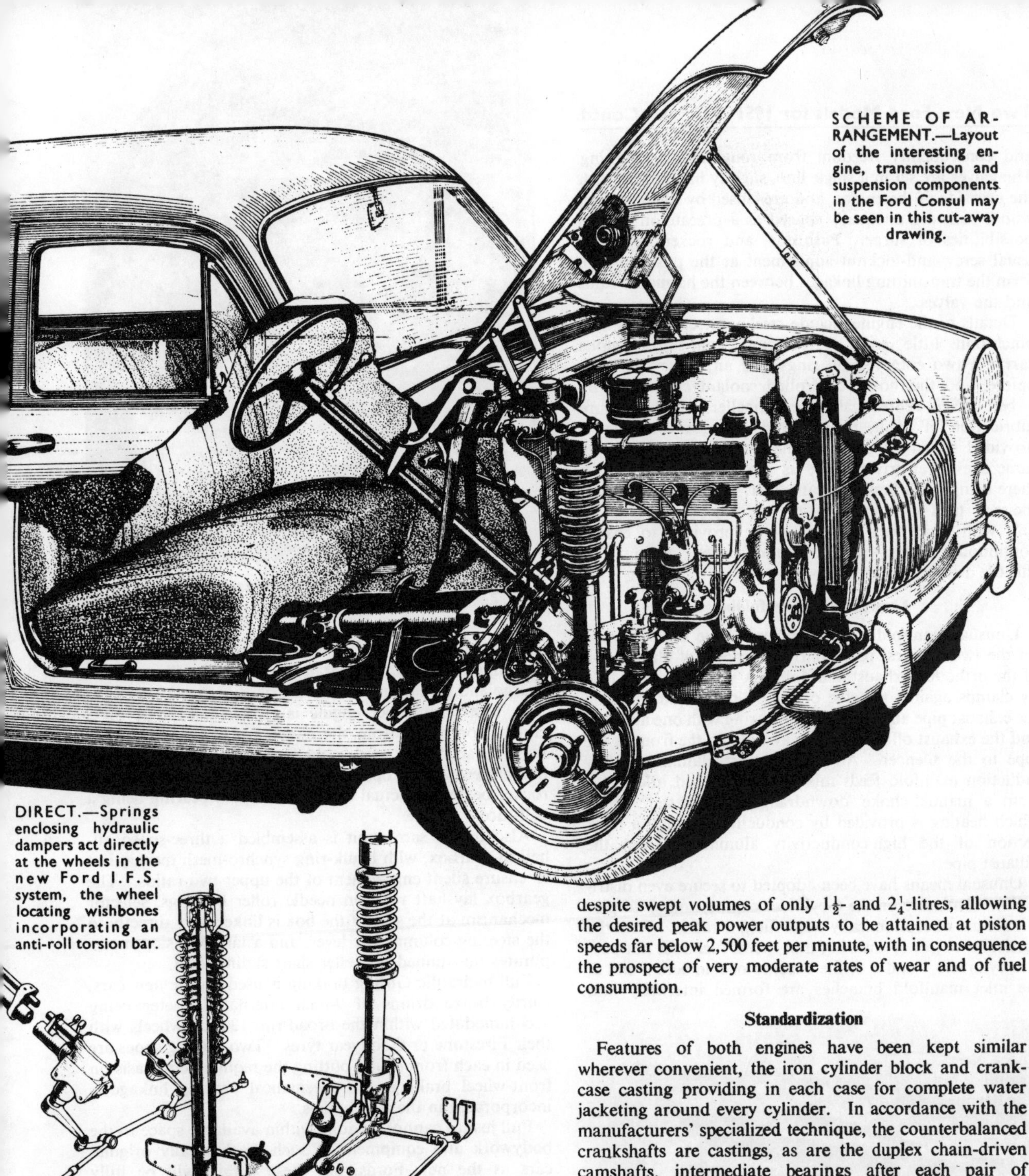

SCHEME OF ARRANGEMENT.—Layout of the interesting engine, transmission and suspension components in the Ford Consul may be seen in this cut-away drawing.

DIRECT.—Springs enclosing hydraulic dampers act directly at the wheels in the new Ford I.F.S. system, the wheel locating wishbones incorporating an anti-roll torsion bar.

Power units are rubber-mounted at three points in the cars, two mounting points alongside the engine being provided on the tubular cross member mentioned in connection with the front suspension, the third at a cross-member passing below the tail of the gearbox. The mounting at the latter point has been specially designed to give controlled flexibility, absorbing any vertical or lateral engine vibration yet providing positive fore-and-aft location of the complete power unit.

Totally new, the 4-cylinder Consul and 6-cylinder Zephyr engines both have identical cylinder dimensions, the stroke in each case being slightly smaller than the bore. Thus, horse-power ratings are 16 and 24 respectively, despite swept volumes of only 1½- and 2¼-litres, allowing the desired peak power outputs to be attained at piston speeds far below 2,500 feet per minute, with in consequence the prospect of very moderate rates of wear and of fuel consumption.

Standardization

Features of both engines have been kept similar wherever convenient, the iron cylinder block and crank-case casting providing in each case for complete water jacketing around every cylinder. In accordance with the manufacturers' specialized technique, the counterbalanced crankshafts are castings, as are the duplex chain-driven camshafts, intermediate bearings after each pair of cylinders being used.

Overhead valvegear is a novelty on a Ford car, but apart from other virtues it provides accessibility suited to this particular installation. Each cylinder has a combustion chamber of wedge form, spread of combustion from a long-reach 14 mm. sparking plug at the base of the wedge towards the apex giving a controlled rate of burning

and consequently freedom from roughness or pinking. The valves are set in a single line, slightly inclined towards the nearside of the engine, and are closed by single springs whose coils are unequally spaced as a precaution against possibilities of surge. Pushrods, and rockers with the usual screw-and-locknut adjustment at the pushrod end, form the transmitting linkages between the harmonic cams and the valves.

Details of the engine include steel connecting rods with pinch-bolt little ends, autothermic split-skirt pistons carrying two compression rings and an oil scraper ring apiece, and thermostat-controlled coolant circulation by a belt-driven centrifugal water impeller. Full pressure lubrication of main, big-end and valvegear bearings is provided by an oil pump in the sump, mounted on the same skew-gear-driven shaft as the ignition distributor, there being a full-flow lubricant filter. Fuel is drawn from the rear tank by a camshaft-driven diaphragm pump, in unit with which is a vacuum pump provided to ensure continuous screen-wiper operation even during full throttle driving.

STEP IN.—Low floor level and comfortable rear seating are Zephyr Six features emphasized by this photograph.

Unorthodox Manifolding

Unusual manifolding arrangements have been adopted on the two engines, particularly in respect of elimination of the orthodox exhaust manifold. A straight pipe is held by clamps against the side of the cylinder head, ports in the exhaust pipe and the head registering with one another, and the exhaust offtake curving away from the front of the pipe to the silencer. An external cast aluminium alloy induction manifold feeds mixture to siamesed inlet ports from a manual-choke downdraught carburetter, below which heating is provided by conduction through a thick section of the high-conductivity aluminium from the exhaust pipe.

Unusual means have been adopted to secure even distribution of mixture on the 6-cylinder engine, and so to permit the use of economically weak fuel/air ratios. With the particular object of ensuring that excess fuel does not pass into cylinders 1 and 6 at the expense of numbers 2 and 5, the inlet manifold branches are formed internally into something in the nature of a screw thread, designed so that in conjunction with the 90-degree bend it will swirl the charge and give even mixture distribution.

TAILPIECE.—Ample accommodation for luggage is provided alongside the spare small-diameter wheel.

In unit with either engine is mounted a single plate clutch, controlled from the clutch pedal by a hydraulic hook-up such as is used in the braking system, fluid from the master cylinder operating a slave piston, push-rod, and rocker beam, the actual clutch withdrawal bearing being a ball-race.

Also in the same unit is assembled a three-speed all-helical gearbox, with baulk-ring synchro-mesh mechanism to ensure silent engagement of the upper two ratios. The gearbox layshaft runs on needle roller bearings, selector mechanism at the side of the box is linked very directly to the steering-column gear lever, and a tail extension incorporates the splined propeller shaft sliding joint.

Full hydraulic Girling braking is used on the new cars, sturdy brake drums of 9-inch internal diameter being accommodated within the broad-rim 13-inch wheels with their Firestone or Goodyear tyres. Two leading shoes are used in each front drum, putting the required emphasis on front-wheel braking, and mechanical parking linkage is incorporated in the rear brakes.

Full justice cannot be done within available space to the bodywork and equipment of such mechanically original cars as the new Fords, but these details will be fully appreciated by Motor Show visitors. Moderate in overall height but with low floor levels, the new Ford bodies make the utmost use of their 64-inch overall width. Prominent internal features are the instruments grouped in front of the driver, the use of transparent plastic control knobs in a manner hitherto associated with expensive luxury cars, and the bench-type tubular-framed adjustable driving seat. Hinged ventilation panels are fitted on the front doors, and fresh air ducts which lead from the front of the car into the body are arranged so that they may be linked to the interior heating system which is an optional extra.

Electrical equipment is planned around a 12-volt generator, inbuilt headlamps and separate parking lamps being provided. Instruments comprise the speedometer, ammeter and fuel contents gauge, plus warning lamps for oil pressure, dynamo charge, traffic signals and headlamp high-beam, and there are ash-trays and sun vizors fitted as standard equipment. Only part of one side of the roomy luggage locker is occupied by the spare wheel and tyre.

Although it is styled for simplicity and ease of production, the Consul is blessed with clean lines and a smooth appearance. An impression of length is given by the low waistline.

DATA FOR THE DRIVER

FORD CONSUL

PRICE (basic), with saloon body, £425, plus £118 16s 2d British purchase tax. Total (in Great Britain), £543 16s 2d.

ENGINE : 15.6 h.p. (R.A.C. rating), 4 cylinders, overhead valves, 79.37 × 76.2 mm, 1,508 c.c. Brake Horse-power : 47 at 4,400 r.p.m. Compression Ratio : 6.8 to 1. Max. Torque : 74 lb ft at 2,400 r.p.m. 14.92 m.p.h. per 1,000 r.p.m. on top gear.

WEIGHT (running trim with 5 gallons fuel) : 21 cwt 2 qr 7 lb (2,415 lb). Front wheels 58 per cent ; rear wheels 42 per cent. LB per C.C. : 1.6. B.H.P. per TON : 43.6.

TYRE SIZE : 5.90—13in on bolt-on steel disc wheels.

TANK CAPACITY : 9 English gallons. Approximate fuel consumption range, 24–28 m.p.g. (11.77–10.08 litres per 100 km).

TURNING CIRCLE : 40ft 6in (L) ; 41ft 0in (R). Steering wheel movement from lock to lock : 2⅔ turns. LIGHTING SET : 12-volt.

MAIN DIMENSIONS : Wheelbase, 8ft 4in. Track, 4ft. 2in (front) ; 4ft 1in (rear). Overall length, 13ft 6.56in ; width, 5ft 4in ; height, 5ft 0¾in. Minimum Ground Clearance, 6.6in.

ACCELERATION

Overall gear ratios	From steady m.p.h. of 10-30 sec	20-40 sec	30-50 sec
4.625 to 1	12.7	13.1	15.1
7.598 to 1	7.6	8.7	—
13.145 to 1	—	—	—

From rest through gears to :—

	sec		sec
30 m.p.h. . .	7.7	60 m.p.h. . .	31.1
50 m.p.h. . .	19.7		

SPEEDS ON GEARS :

(by Electric Speedometer)	M.p.h. (normal and max)	K.p.h. (normal and max)
1st	20—28	32—45
2nd	40—49	64—79
Top	75	121

Speedometer correction by Electric Speedometer:—

Car Speedometer		Electric Speedometer m.p.h.
10	=	9.5
20	=	19.0
30	=	28.0
40	=	37.5
50	=	47.0
60	=	56.0
70	=	65.0
80	=	74.0
81	=	75.0

WEATHER: *Dry* ; *wind fresh to light.*

Acceleration figures are the means of several runs in opposite directions.

Described in "The Autocar" of October 20, 1950.

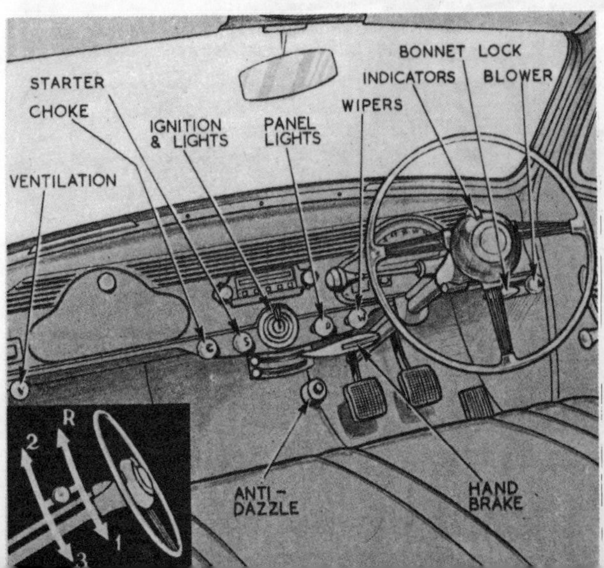

FORD CONSUL SALOON

IT was a considerable motoring event last autumn when the Ford company introduced two entirely new models from Dagenham. As will be remembered, the interest taken in them at the London Show last October was enormous, for so obviously did the four-cylinder Consul and its larger companion, the six-cylinder Zephyr, show that Ford were offering thoroughly up-to-date cars in the medium range of size, and at prices, in the prevailing conditions, of an order always associated with this vast organization.

The new cars are a complete breakaway and bear no resemblance in appearance and general styling, or in mechanical design, to the previous products. Indeed, they have a number of unusual and advanced features of design, especially as regards the front suspension; also they have the innovation, for Ford, of overhead valves. The Consul has come first into production and it is this model which *The Autocar* has had the first opportunity of sampling extensively in the variety of conditions applied to cars undergoing Road Test.

First impressions are of particular importance when a much-heralded car of this significance from a major factory is sampled. No experience is more interesting in the life of anyone who spends much time analysing the road behaviour of different makes of car. First impressions of the Consul are that the engine is smooth and lively and satisfactorily quiet, and that the car is of an overall size handy in city traffic and on narrow and crowded roads, has light, accurate steering and hydraulic brakes that give plenty of power, and that it has good stability. It is one of the outstanding cars produced since the war in the popular class and has handling qualities that would be acceptable on a car of any price.

Thinking of the Consul alone for the moment, Ford's have undoubtedly produced an economical car in the 1½-litre class which is capable of doing everything that the ordinary motorist requires, and of doing it more than adequately, in a way which has appeal also to an experienced driver who can take the broad view contributed to by knowledge of contemporary models. Incidentally, it is not apt to say of the Consul by way of designation that the engine is of a size which used usually to be known as "12 h.p.," for, although it is of a litreage that frequently corresponded to that old-style rating, full advantage has been taken of modern ideas in the design in regard to limitation of piston speed, and the bore is appreciably greater than the stroke. If the old horse-power rating formula still applied in Great Britain the dimensions would give a figure of 15.6 h.p.

With its flush sides running through from front to rear.

The deep windscreen and shallow grille produce a good-looking front view. A wide curved windscreen and a large rear window provide good all-round visibility.

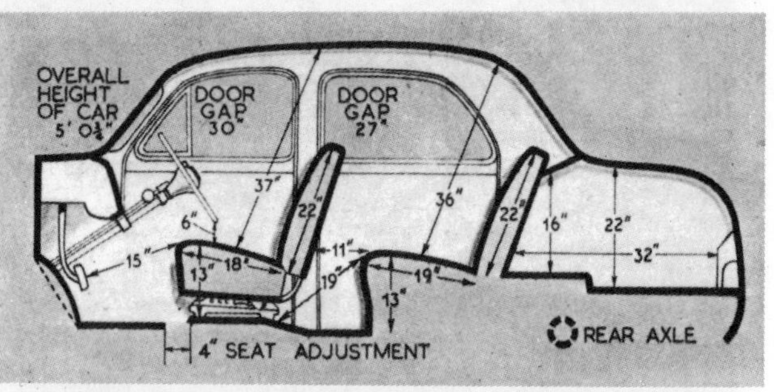

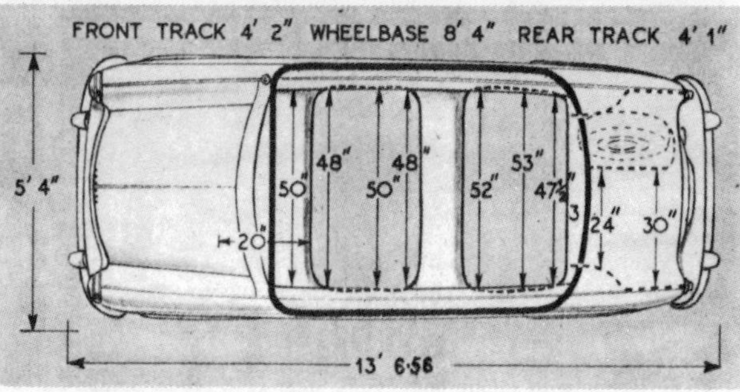

Measurements in these scale body diagrams are taken with the driving seat in the central position of fore and aft adjustment and with the seat cushions uncompressed.

the Consul is a smart car entirely in the modern style. It is of integral construction, that is, without a separate chassis frame, and clearly it is a conception which looks sufficiently far ahead to remain basically settled for a considerable time.

A comfortable speed to maintain when in a hurry is a genuine 55 m.p.h., and there is no increase of mechanical effort or noise at 60 to 65 m.p.h. sufficient to impose restriction on the faster type of driver when he has a suitable road; all-out maximum speed is higher than might have been expected. Such speeds are not everyone's taste, but it is necessary to lay stress on the behaviour of cars at the higher speeds, for at, say, 40-45 m.p.h. there is relatively little difference between any number of cars of broadly the same type. Users of the Consul in a hurry on business will find that there is plenty of reserve performance and that average speeds of the order of 40 m.p.h. are within its easy reach without any feeling of insecurity, for the car rides extremely well and has powerful brakes.

With a ratio which strikes a compromise between low and high for the size of engine, the engine pulls very well on top gear. There is, however, everything against letting it do so beyond certain limits, in view of the fact that there is a useful middle ratio in the three-speed box, engaged by a steering-column gear lever which, because the linkage is simple, works extremely well with only finger pressure and which gives rise to no uncertainty about the engagement of any gear. Long-legged drivers may find the gear lever touching the left leg in top gear position, a matter that would be avoided if the assembly were moved slightly on the column. In the hands of a driver of some experience a given car with four speeds will always possess advantages, but to the great majority of drivers, and to all in some degree, three speeds are adequate when, as with the Consul, the power-to-weight ratio is good. Second gear will take it over a 1 in 6 hill, given a clear run, at a minimum of 25 m.p.h., and there is a reserve of power on this ratio that permits a slow approach to be made to an appreciable gradient, or enables a baulk on such a gradient to be handled, without any question of first gear being required.

One of the unusual features of the new Fords is the hydraulic operation, through the pedal, of the clutch withdrawal. There is a fluid reservoir, accessible under the bonnet, common to the clutch and brake master cylinders. A slightly eager start from rest was experienced at first acquaintance, but after a time this impression disappeared.

The behaviour of the suspension is observed with particular interest in view of its unusual character, coil springs being used but without the usual double wishbone assembly. An anti-roll bar acts as the two forward arms of a pair of wishbones, whilst each front wheel and brake assembly is attached to a tubular king-pin forming the cylinder of a direct-acting hydraulic damper. The suspension permits very little roll unless extreme methods, outside ordinary reckoning, are adopted, and the riding is level on average surfaces, whilst severe shock is absorbed most efficiently. Vertical motion is very satisfactorily small.

Without being of that pattern, the steering gives the impression of possessing advantages associated with rack

A simple trim style and the absence of excessive decoration and plating are evident in the interior views. The clutch and brake pedals are pivoted from the top, and an "organ pedal" type of throttle is used. The seat adjustment control is convenient at the side of the cushion. The low floor level necessitates a considerable "step down" when entering the car. Substantial "pulls" are fitted on the inside of all doors, which remain firmly in the fully open position under the control of check devices. There is a useful locker in the facia and three ashtrays are fitted.

The locker *lid*—rather than " door " on the average European car—and the spare wheel position show perhaps the American influence in design. Ford ingenuity is well to the fore in the arrangement of the number plate illumination, which can be seen on the lower edge of the lid.

The four-light body is balanced by a " bonnet-like " luggage locker, and the clean exterior is not spoiled by exterior hinges that have become " popular " since the war.

and pinion steering in being finger light, yet definite, and leaving no room for doubt as to whether the steering wheel is connected to the road wheels by positive mechanical means. It is medium/high geared steering and earns high praise for the qualities already enumerated and for the absence of transmitted road-wheel shock. The hydraulically operated brakes give one immediately the suggestion of being adequate but not unusually powerful. As miles are covered and various conditions encountered it is realized that they are deceptive because of a very light pedal operation, and that the right degree of deceleration is produced as required with increased pedal pressure. Slight limitation of the steering lock is noticed for the size of car, in circumstances where comparison exists.

For the driver perhaps a more vertical back rest of the one-piece front seat would give better support, its angle of rearward inclination not being particularly liked by one average-height driver responsible for these impressions, in conjunction with a cushion which is higher at the front edge than he happens to favour, the combination tending to give a "folded-in-the-middle" feeling. Apart from fairly wide, though not awkwardly obtrusive, windscreen pillars, outward vision is good and the central ridge of the left wing is seen by way of useful guidance in a right-hand drive car. The spring-spoked steering wheel is at a comfortable angle.

Instruments are simplified to speedometer, petrol gauge and ammeter for the 12-volt electrical system, an oil-pressure gauge being replaced by a warning light; a clock is not fitted. The dials are well illuminated at night without causing glare. An excellent view is given by the driving mirror, which is tinted to reduce glare from following lights at night in the now usual absence of a rear window blind. The size of the rear window itself not only helps to make the interior light and spacious feeling, but also facilitates reversing in congested places. In the facia in front of the passenger there is a small cubby hole provided with a lid, but no other provision is made for carrying impedimenta. Two sun vizors are fitted.

Ventilator panels are fitted in the front doors and they can be swung round to induce forced ventilation. There is built-in provision for further controlled ventilation and for windscreen demisting, an interior heater unit being available at £12, plus purchase tax on the home market; this was fitted on the car tested. Leg room is good in spite of intrusion of the wheel arches into the front compartment. Interior finish is plain but adequate and serviceable, and the seating is within the wheelbase, a point of marked benefit, of course, as regards riding in the rear seats. The luggage locker is of very useful size, utilizing the full width of the car.

A quick, no-fuss getaway is made from cold. The choke control has several positions at which it can be locked from full rich back to normal, and, the engine having fired instantly during the test, when it stood in the open overnight on several occasions, the full-rich position could quickly be dispensed with, and in a very short distance the control be returned to normal, the car getting under way without spluttering or requiring to be coddled.

The horn note is slightly unusual in character, not unpleasingly, and is effective. The head lamp beam is not exceptional but is adequate. Double-filament bulbs are fitted in both lamps for anti-dazzle purposes. There is much common sense—a quality markedly displayed throughout—about the suction-operated twin screenwipers. They are a little audible, but they wipe particularly clean, with a strong action, on the curved glass, even in the conditions of spray from a wet road when it is not actually raining; they spare the battery, and are free from the disadvantage of early suction-driven wipers that dried up at a wide throttle opening. On the Ford they are operated from a vacuum pump in conjunction with the camshaft-driven fuel pump, and not from the induction manifold, as this pattern of wiper used to be.

On the left-hand side above the heater inlet duct are the fluid reservoir and twin master cylinders for the clutch and brake mechanisms. The horn is fitted on the under side of the bonnet, which is held open by a self-locking strut. The radiator is surrounded by valances, which are in fact part of the front " frame " assembly.

The Motor Continental Road Test No. 6C/51

Make: Ford. **Type:** Zephyr Six.

Makers: Ford Motor Co. Ltd., Dagenham, Essex.

Dimensions and Seating

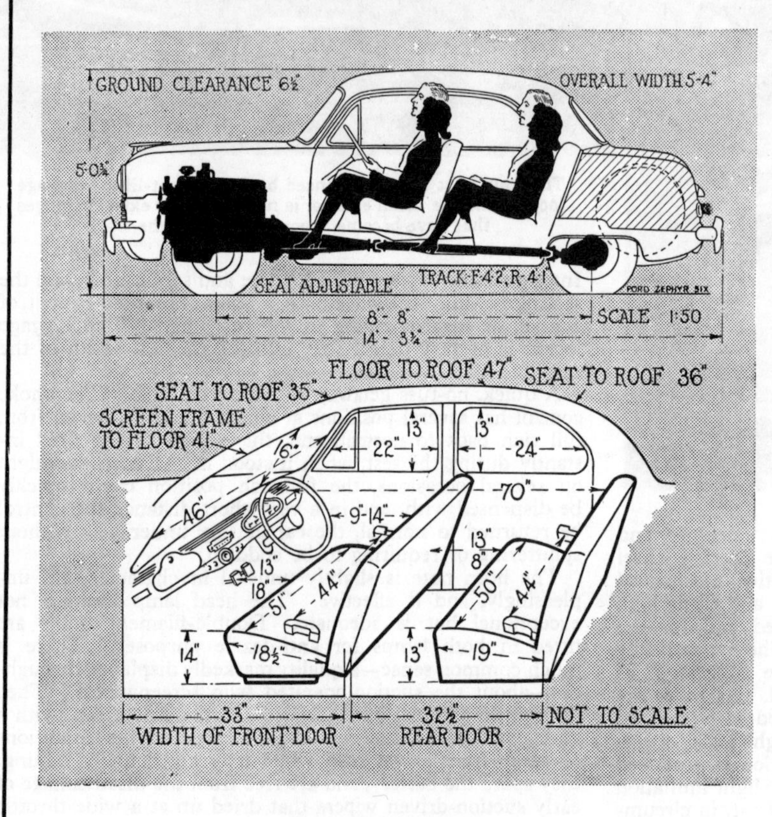

Test Conditions

Light breeze, dry ; smooth tarmac and concrete surfaces. Pool petrol.

Test Data

ACCELERATION TIMES on Two Upper Ratios

	Top	2nd
10–30 m.p.h.	7.5 secs.	4.9 secs.
20–40 m.p.h.	8.2 secs.	5.7 secs.
30–50 m.p.h.	9.4 secs.	8.1 secs
40–60 m.p.h.	11.8 secs.	—
50–70 m.p.h.	16.0 secs.	—

ACCELERATION TIMES Through Gears

0–30 m.p.h.	5.0 secs.
0–40 m.p.h.	8.8 secs.
0–50 m.p.h.	14.0 secs.
0–60 m.p.h.	20.2 secs.
0–70 m.p.h.	29.9 secs.
Standing quarter-mile	21.8 secs.

MAXIMUM SPEEDS

Flying Quarter-mile

Mean of four opposite runs	79.8 m.p.h
Best time equals	81.1 m.p.h

Speed in Gears

Max. speed in 2nd gear	50 m.p.h.
Max. speed in 1st gear	29 m.p.h.

FUEL CONSUMPTION

24½ m.p.g. at constant 30 m.p.h.
24 m.p.g. at constant 40 m.p.h.
23½ m.p.g. at constant 50 m.p.h.
21 m.p.g. at constant 60 m.p.h.
Overall consumption for 202 miles, 8½ gallons, equals 23.7 m.p.g.

WEIGHT

Unladen kerb weight	22¾ cwt.
Front/rear weight distribution	58/42
Weight laden as tested	26½ cwt

INSTRUMENTS

Speedometer at 30 m.p.h.	Accurate
Speedometer at 60 m.p.h.	1% slow
Distance recorder	3% fast

HILL CLIMBING (at steady speeds)

Max. top-gear speed on 1 in 20	69 m.p.h.
Max. top-gear speed on 1 in 15	64 m.p.h.
Max. top-gear speed on 1 in 10	53 m.p.h.
Max. gradient on top gear	1 in 8.2 (Tapley 270 lb./ton)
Max. gradient on 2nd gear	1 in 5.1 (Tapley 435 lb./ton)

BRAKES at 30 m.p.h.

0.91 g. retardation (=33.0 ft. stopping distance) with 140 lb. pedal pressure.
0.63 g. retardation (=47.5 ft. stopping distance) with 75 lb. pedal pressure.
0.48 g. retardation (=62.5 ft. stopping distance) with 50 lb. pedal pressure.
0.29 g. retardation (=103.5 ft. stopping distance) with 25 lb. pedal pressure.

In Brief

Price £540 10s. plus purchase tax £301 15s. 7d. equals £842 5s. 7d. (As tested, with leather upholstery, heater and radio= £896 14s. 5d.)

Capacity	2,262 c.c.
Unladen kerb weight	22¾ cwt.
Fuel consumption	23.7 m.p.g.
Maximum speed	79.8 m.p.h.
Maximum speed on 1 in 20 gradient	69 m.p.h.
Maximum top gear gradient	1 in 8.2
Acceleration	
10-30 m.p.h. in top	7.5 secs.
0-50 m.p.h. through gears	14.0 secs.

Gearing: 16.3 m.p.h. in top at 1,000 r.p.m. 81.5 m.p.h. at 2,500 ft. per min. piston speed.

Specification

Engine	
Cylinders	6
Bore	79.37 mm.
Stroke	76.2 mm.
Cubic capacity	2,262 c.c.
Piston area	46.0 sq. in.
Valves	Pushrod O.H.V.
Compression ratio	6.8/1
Max. power	68 b.h.p.
at	4,000 r.p.m.
Piston speed at max. b.h.p.	2,000 ft. per min.
Carburetter	30 mm. downdraught
Ignition	12-volt coil
Sparking plugs	Champion NA8, 14 mm.
Fuel pump	AC mechanical
Oil filter	AC full-flow
Transmission	
Clutch	8-in. single-plate
Top gear (s/m)	4.375
2nd gear (s/m)	7.187
1st gear	16.875
Propeller shaft	Open
Final drive	Hypoid bevel
Chassis	
Brakes	Girling hydraulic
Brake-drum diameter	9 ins.
Friction lining area	121 sq. ins.
Suspension :	
Front	Coil spring i.f.s.
Rear	Semi-elliptic
Shock absorbers :	
Front	Direct acting
Rear	Double acting piston
Tyres	6.40×13
Steering	
Steering gear	Burman
Turning circle	43 ft.
Turns of steering wheel, lock to lock	2¾
Performance factors (at laden weight as tested)	
Piston area, sq. ins. per ton	35.0
Brake lining area, sq. ins. per ton	92
Specific displacement, litres per ton/mile	3,170

Fully described in "The Motor," October 18, 1950.

Maintenance

Fuel tank : 9 Imp. gallons. **Sump :** S.A.E. 20 or 20 W. 8 pints (plus 1 pint for filter). **Gearbox :** 2 Imp. pints S.A.E. 80 E.P. **Rear axle :** 2½ Imp. pints S.A.E. 90 Hypoid. **Steering gear :** S.A.E. 80 E.P. **Radiator :** 22 pints (2 drain taps). **Chassis lubrication :** By grease gun every 1,000 miles to 12 grease points plus 2 oil points (universals). **Ignition timing :** 11 degrees before T.D.C. (static). **Spark plug gap :** 0.032 in. **Contact breaker gap :** 0.012–0.014 in. **Valve timing :** I.O. 17 degrees B.T.D.C., I.C. 51 degrees A.B.D.C. ; E.O. 49 degrees B.B.D.C., E.C. 19 degrees A.T.D.C. **Tappet clearances** (hot) : Inlet 0.014 in., exhaust 0.014 in. **Front wheel toe-in :** $\frac{1}{16}$ in. to $\frac{1}{8}$ in. **Camber angle :** $-\frac{1}{2}$ degree to $+\frac{1}{2}$ degree. **Castor angle :** Zero to $+1\frac{1}{2}$ degrees. **Tyre pressures :** Front and rear 24 lb. (minimum). **Brake (and clutch control) fluid :** Girling crimson. **Battery :** 12 v. 45 amp./hr. **Lamp bulbs :** All 12 v. except bulbs in instrument cluster. Head lamps 42/36 w. ; side and rear number plate bulbs 6 w. ; rear and stop lamp 6/24 w. ; indicator and interior lamp bulbs (festoon) 6 w. ; instrument panel bulbs and warning lights for direction indicators, ignition, head lamp beam and oil pressure all 16–18 v 3 w

Ref. B/23/51

The FORD Zephyr Six Saloon

A High Performance Saloon with Fine Roadholding

HANDSOME—Clean, attractive lines are given an increased appeal by the sensible lack of ostentatious decoration.

THE Ford Motor Co. have already two historic innovations to their credit; the T model, introduced in 1908, was the first car which could be owned, driven and maintained by unskilled persons driving over any road conditions in any part of the world; the introduction of the V8 power unit in 1932 marked the first occasion on which a car having a power/weight ratio of some 70 h.p./ton and maximum speed of *circa* 80 m.p.h. was built in very large quantities and offered to the U.S. public at the exceedingly modest price of £125 at the then current exchange rate. Thus, the Ford Co. were the first to give motoring to the masses and the first to give performance to the many.

The new Zephyr which emanates from the British Dagenham section of the Ford group of companies is yet another first, for it allies the high performance and (by modern standards) modest price characteristic of the V8, with a general roadworthiness and stability which virtually take it out of the family motoring class and put it into the sports-car category in a manner obviously remote from the original intention of the designers.

The outstanding performance of the car is very largely based upon the exceptional piston area following from the use of a somewhat larger bore than stroke. This makes it possible to gear the car so that the engine is doing some 5,000 r.p.m. at a maximum speed without, however, exceeding 2,500 ft./min piston speeds. Hence, although the power/weight ratio can be equalled or even exceeded on a number of rival designs, the car has an exceptionally high displacement factor expressed in litres per ton mile, whilst added to these natural advantages it is obvious that the engineers have produced a remarkably high b.m.e.p. in the engine over the lower part of the revolution scale.

The net result of all this is a 10-30 m.p.h. top gear acceleration time which has been bettered by only one closed car in the post war series of "Motor" Road Tests, and this one having an engine nearly twice the Zephyr's capacity and selling at nearly twice the Zephyr's price.

The second gear acceleration corresponding to this remarkable top gear performance is so great that wheel-spin can readily be induced on dry roads—in fact on wet roads one can get wheel-spin in top gear if the throttle is not handled with a certain degree of discretion.

It will be seen from the data figures that the acceleration is strongly maintained up to over 60 m.p.h. and an ability to climb a 1 in 20 gradient at nearly 70 m.p.h. makes it possible to put up some highly remarkable average speeds.

The actual maximum of the car, although far above that used by most motorists, is somewhat lower than might be expected and this is undoubtedly due to the choice of gearing and those features of engine design which determine the very fine bottom-end performance. Moreover, the engine turns with really electric motor smoothness up to 4,000 r.p.m. and at, say, 65 m.p.h. becomes slightly more detectable with increasing speed so that although there is little doubt that it will come to no mechanical harm if held at even 75 m.p.h., on long runs, the driver does in fact find that he settles the car down at some 10 m.p.h. below the maximum.

One aspect of the choice of high crankshaft, and low piston, speed calls for comment. At only a fraction over the true maximum road speed, i.e., at 80 m.p.h. the engine suffers from valve bounce in top gear and should anyone not recognise the source of a slight clatter when driving all-out over some distance damage might result or rapid wear be occasioned.

Accurate and Sensitive

There is a high proportion of weight on the front wheels compared to the back and this results in a decisive break-away at the rear end if corners are taken too fast but in the general run of motoring definite under-steering characteristics are discerned and, particularly on long high-speed bends, the car handles in such a way as to instil very great confidence, this being reinforced by the accurate and sensitive steering gear and the potent braking system which, however, needs rather higher pedal pressure than often obtains today.

One might suppose that the high acceleration and cruising speed, giving together remarkable average speeds, would result in rather poor fuel consumption and such suspicion may seem to be confirmed by the figures given for steady running conditions, as set out in the data sheet. Nevertheless, the overall consumption of the car is far better than one might expect and strictly comparable with other cars in the same class with distinctly lower performance.

If we have stressed the speed and power of this car it is only because they are so

STEP IN—Forward-hinged doors give access to the roomy interior, which derives a sense of lightness from the large window areas, the windscreen being notably deep and wide. All seating is comfortably set within the wheelbase.

CAPACIOUS—The boot takes a far greater quantity of luggage than first appearances suggest. A two-gallon petrol can here gives an idea of its depth, and the shape of the boot lid aids in the carriage of bulky cases.

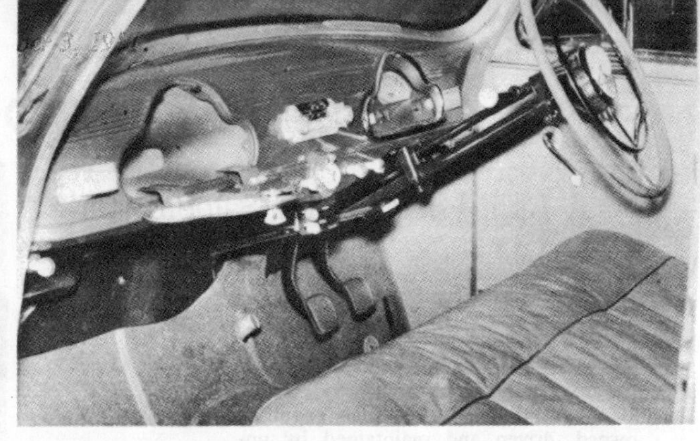

CONTROL POINT—(*Right*) Basic instruments are grouped in front of the driver, a shaped recess being matched on the passenger's side by a cubby hole. Top-hinged pedals, radio and heater controls are features visible here.

Ford Zephyr - - Contd.

exceptional as to overlay in one's mind many of the other excellent features in the design and construction. Chief amongst these must be accounted the clean appearance and fine styling of the car, the simple lines of which constitute a three dimensional reproach to those who consider that eye appeal can only be obtained by excessive ornament. The photographs and drawings show that the car is low-built as a whole and also that it has a particularly low bonnet line and that this in conjunction with the wide curved windscreen gives excellent forward visibility, both sides of the car (one can no longer say mudguards) being very fully within the driver's view. Corresponding therewith is a similar rear view through a large back window.

At the cost of putting the front screen rather a long way from the driver's eye a wide front door gives unusual ease of entry and exit, but actual leg room within the body is limited, for the large diameter of the six cylinders results in a somewhat long engine whilst rear seat passengers are placed well forward of the rear axle centre.

A driver used to the more old-fashioned type of car may at first sight object to the windscreen position, but although there may be exceptional conditions in which the distance between the driver's eye and the screen glass may be a disadvantage, one soon becomes used to this arrangement in ordinary motoring.

There are a number of other features in the body design and equipment which are more legitimately open to criticism. Chief among these is the rather limited locker space in the facia panel and the absence of any alternative carrying capacity in the shape of a facia undertray or sidepockets. There is, however, a generous shelf behind the rear seat, and the luggage boot, which externally looks somewhat on the small side, is in fact one of the most capacious ever provided on a car of this size and although, the spare wheel is contained therein, a party of four, using a quantity of moderate-sized suitcases, could easily tour the Continent without recourse to a roof rack. The fuel filler, incidentally, projects horizontally from the back of the car and this makes it almost impossible to refuel from a can without an extension.

Essential Instruments

The instruments on the panel are limited to a speedometer, fuel gauge and (very welcome) ammeter, with lights showing charge and satisfactory oil pressure, but although both the driver and front passenger have armrests on the doors there are no centre armrests for either the front or back seats.

Finally, one may register surprise that the windscreen wipers continue to be pneumatically driven for although this is undoubtedly a reliable method it suffers from a change of speed with throttle opening which is irritating, and to a degree of noise on part throttle which many must consider maddening. The range of speed of these wipers can, however, be adjusted by the driver to suit conditions.

In addition to these detail points one observes that the excellent road holding is matched by a certain stiffness at low speeds on rough roads although the combined body and chassis construction appears to be exceedingly solid with very little shock in the structure. Some drumming is also apparent on rough roads.

A slight whistling noise emanates at high speed from the highly effective body ventilation system to which was added on the car tested a heater, de-mister, and defrosting element complete with supplementary electric fan.

The body is both water and air-tight, and although it was impossible to carry out tests at low temperatures one has the feeling that the car would survive a test of this kind particularly well. Taking the other extreme—an extra fresh air supply can be admitted into the body as needed by turning the triangular ventilators in the front windows through more than a right angle.

Low Noise Level

The car also carried supplementary equipment in the shape of an Ekco press-button radio set which gave good quality reproduction through a loud speaker centrally mounted above the windscreen, the set drawing electrical impulses from a mast which could be extended to more than usual height. On normal English road surfaces this set could be listened to with pleasure at speeds between 65 and 70 m.p.h.—a tribute not only to the manufacturers of the set, but also to the comparatively low road, mechanical and wind noises produced by the car.

The general accessibility of the engine and battery appears to be good and the finish and furnishings simple in style and of a quality which is perhaps above that which one might expect in view of the price of the car.

This aspect of price should indeed be borne constantly in mind. No one should permit the prominence of a few detail defects in minor items of the Zephyr specification to outweigh the remarkable achievement represented by the vehicle as a whole.

To sum up, it will carry six people in comfort and therefore with great room to spare; it will out-accelerate any car in its class and will maintain station with anything outside the large engine luxury, or sports-car, category. It can be cruised at between 70/75 m.p.h. and is in addition a very safe and pleasurable car to drive. Finally, one must say that it represents one of the finest examples of value for money to be found today not only amongst English but also amongst world production.

POWER PACK—The wide, single-piece alligator bonnet gives access to the engine and accessories. Note the twin horns mounted on the bonnet underside.

FORD ZEPHYR-SIX TEST REPORT

$2\frac{1}{4}$-litre Saloon with "Over-Square" Engine Provides Effortless and Comfortable 70 m.p.h. Travel and is a Pleasure to Drive

HAVING taken a Ford Consul up to Charterhall and back for the excellent racing at Scotland's first International Race Meeting last October, MOTOR SPORT recently had the pleasure of doing a rapid day-tour into Wales with a Ford Zephyr-Six.

On the latter occasion we took no especial note of the average speeds accomplished, for such observations are of interest only in respect of extremely fast cars over uninterrupted journeys. But what the Ford accomplished was impressive in a different way, for here was an economical, comfortable saloon of compact dimensions and modest price, which packed in an enormous amount of travel without tiring itself or its occupants.

Leaving Lewknor, on the London side of Oxford, at 8 a.m., we were able to motor to Dinas Powis, search for and locate an early Lagonda, spend half-an-hour examining this hedgerow discovery, drive over snow-covered roads through the Rhondda Valley, lunch, carry our exploration on to Presteign, of V.S.C.C. rally memory, attempt some cross-country stuff where we stuck for something like an hour in a drift, take a very leisurely meal at the "Radnorshire Arms," and drive back to our starting point over slippery roads without hurrying, arriving by 10.15 p.m.

A car which can so usefully cover the ground is the sort of transport which 75 per cent. at least of the world's car-buyers require. The three enthusiasts who made this day tour (which embraced that excellent free entertainment, the aerial voyage across the Usk on the Newport Transporter Bridge and some fine motoring along deserted mountain roads with the sun glinting on snow-capped peaks) were all warm in their appreciation of this Zephyr-Six.

Apart from its easy running and comfortable riding, this latest Ford is notably economical, giving 25 m.p.g. It is also a quite good cross-country proposition, its excellent ground-clearance and high-set exhaust system making light of the "back o' beyond," as negotiation of a deep water-splash emphasised. In this connection the back wheels carry less load than is usual and consequently tend to break-away rather early on corners or to spin on slippery hills, but no doubt this could be countered by carrying ballast in the boot or fitting chains. Apart from this tendency to lose adhesion, the three-speed gearbox has well-chosen ratios and the engine sufficient power to render easy a restart in the 12.62 to 1 bottom gear on steep gradients. Full power in first gear, however, brought in very noticeable rear-wheel judder.

On long journeys this Zephyr-Six excells, for it cruises at 70-75 m.p.h. with an easy stride, accelerating powerfully and taking ordinary hills in the highest ratio with little diminution of speed. The driver has an excellent view ahead through the broad screen and can see both front mudguards. It takes him a little time to realise that the wheels are set quite appreciably inboard, so that kerbs can be clipped closer than at first appears prudent. The 17-in. steering wheel is set sensibly far from the facia and it controls really good steering, quick and smooth, working "against the castor" sufficiently to remain taut and tell you where the front wheels are, yet not functioning at all heavily. There is rather vigorous castor action, no return-motion and only very slight vibration is transmitted, and the gearing, at $2\frac{1}{2}$ turns lock-to-lock, is an excellent compromise which only the vintage-car addict might term a trifle low-ratio. The wheel carries direction-indicator switch control and a half-horn-ring, the latter, turning with the wheel, apt to be lost to the finger-tip at moments when it is most needed.

The cornering tendency is towards under-steer and the Zephyr-Six handles very well, instilling great confidence over wet, even icy, roads, and contributing greatly to the pleasure derived by a keen driver. The aforesaid early rear-end break-away can be adequately met under most circumstances by the light, sensitive steering, for which the Ford engineers deserve full marks. Moreover, the Zephyr rides on a reasonably level keel, and any rolling which does take place can be met by the responsive, smooth steering.

The Zephyr rides the rough stuff with its 13-in. wheels bouncing about furiously but its occupants isolated from road shock, which is exactly as it should be. The suspension is, additionally, efficiently damped, so that hump-back bridges and road-opener's gulleys have no effect on control. Full marks again! In Monmouthshire we met a long stretch of road which would have made M.I.R.A. jealous but the Ford rode it comfortably and securely. There is some up and down motion but normally the car rides smoothly.

A centre arm-rest for the front seat would be useful; a leather-upholstered bench seat particularly calls for this aid to dignified passengering. The back seat is so provided and the doors have arm-rests.

At speeds above 60 m.p.h. a grumbling noise intrudes, which might be a source of appreciable irritation on Continental journeys, where such a speed would be held for hours on end. It was accompanied by some floor vibration and would seem to be the effect of certain road surfaces on small tyres, magnified by the one-piece construction. There was also a good deal of wind-noise, but the Ekco radio, occupying some of the otherwise extremely generous area of the under-facia parcels shelf, competed successfully. Otherwise the car is extremely quiet and the six-cylinder, 68-b.h.p., o.h.v. engine very smooth. The latter starts promptly, does not run-on, and suffers only subdued "pinking" on Pool. It called for no water and little oil after 600 miles. Valve bounce comes in at 4,000 r.p.m., equal to 28 m.p.h. in first, 50 m.p.h. in second gear.

The bench seats are comfortable but leg-room is a thought restricted by the length of the big-bore six-cylinder engine and the good feature of a rear seat within the wheelbase. A medium-height human would not remark on this, however, but a taller occupant also finds some restriction of head-room in the back seat. The front seat adjusts easily, the rear windows please by winding fully down if required, most of the expected "mod. cons." are present and the heater became effective after we had used the cover of the Ford catalogue to blank-off part of the radiator. In doing this we discovered that the bonnet-release handle needed a great deal of force to make it function, that dip-stick, battery, etc., are usefully accessible, but thereafter we felt the lack of a coolant thermometer or oil-pressure gauge. The pull-out knobs on the facia, one sufficing for the lamps control, are adequate, once you realise that "B" means heater-fan, "V" ventilator and "Hood" is the bonnet-release. The protruding-cowl instrument cluster before the driver consists of speedometer-cum-odometer (*sans* trip and with no "tenths"), ammeter, fuel gauge, and warning lights for ignition, oil, indicators (tiny arrows) and headlamps beam. Ignition key and starter button are on opposite ends of the cluster. The fuel gauge is usefully pessimistic but its dial has no readings save four spots, those indicating empty and quarter-full being very close-spaced.

The hanging pedals work well, likewise the spoon-type accelerator, but the hydraulically-actuated clutch is rather heavy to hold down, and somewhat sensitive to engage with a long pedal travel. The 9-in. Girling brakes, if calling for determined pedal pressure, can be dismissed with full marks, and the central pull-out-from-facia handbrake is good of its kind and holds the car like a rock, but is too close to the heater control quadrant. There are twin sun vizors with a neat radio loudspeaker between them. The luggage locker is particularly spacious and although the spare wheel lives therein it has the decency to stand upright. The locker lid is easy to lift, and locks.

The steering-column gear-change functions adequately, with useful synchromesh, and the rigid lever is free from that troublesome

CONTINUED ON PAGE 66

ZEPHYR SIX.—Roomy and economical, the Ford Zephyr is a fast and comfortable means of transport, pleasant to drive, and selling at a competitive price.

Towards the Century

AN INTERIM REPORT ON SOME EXPERIMENTAL WORK WITH A FORD ZEPHYR

By Laurence Pomeroy
F.R.S.A., M.S.A.E.

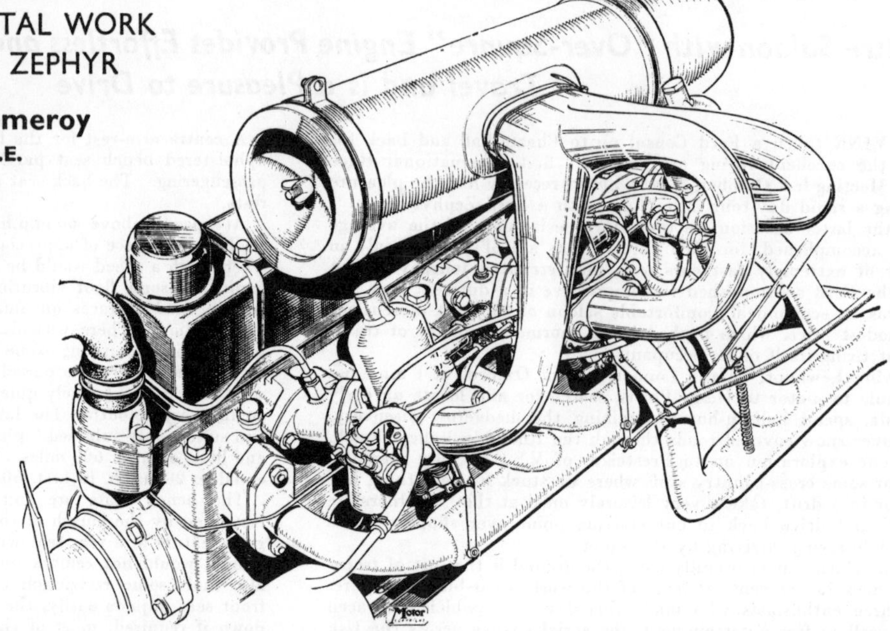

NINE-PIPER.— This drawing of the revised inlet and exhaust manifolding on the Ford Zephyr shows that air is received by three S.U. carburetters from a Vokes air silencer and cleaner through a three-branch light-alloy pipe. Fuel and air are delivered by three tapering stub pipes to the siamesed inlet ports. Each of the six exhaust ports has a separate offtake pipe, the front three and back three cylinders being coupled so that there are two entirely separate exhaust systems each with its own manifold.

ON a number of occasions I have batted for several hours, but I must confess that I have never made as many as 20 runs. Others, more prolific in this respect, have told me that one of the most difficult periods in the game is when one has made 94, or more, and the magic century can well depend on a single hit.

This seems a satisfactory analogue with the performance of motorcars, for 90 m.p.h. is not too difficult to reach, but a genuine " 100 " is a rare achievement. It is, nevertheless, a speed that I am not without hope of attaining on a considerably modified Ford Zephyr. This has aroused some interest since it came on to the road a few weeks ago, and although a considerable amount of development work remains to be done, an interim report may interest owners of this inherently attractive car.

From 1927 until 1937 I drove almost nothing but test and experimental vehicles, and although the performance of these was often startling the possibility of arriving at one's destination without trouble was small; indeed there was always the chance that one would not arrive at all. It came, therefore, as something of a refreshing change to join *The Motor* in 1937 and to receive for the first time in my life not only a steady monthly income, but also completely reliable transport in the form of a 2-litre M.G. In the immediate post-war period I have continued to enjoy both of these boons, in that the Triumphs (Roadster and Mayflower) and Morris Minor, of which I have enjoyed the use, have been 100% standard cars and have given very nearly 100% faithful service. Just, however, as the criminal returns to the scene of his crime, and men (and perhaps women, too) return to thoughts of their first love, so has the desire to revert to experimental work been increasingly difficult to suppress. Before, however, embarking upon the modification of a complete car two conditions had to be met. These were that the subject of the work had to be basically sound, and worth the extra effort, and (perhaps more difficult) of initially so low a price that the overall cost would not be absurd.

Experience with the Ford Zephyr showed that it satisfied both of these conditions and after gaining experience over a few thousand miles I was able to put in hand a series of modifications aimed in the first instance at raising the peak engine r.p.m. and maximum engine power.

The standard Zephyr has a power peak of 4,400 r.p.m. at which it develops some 64 h.p., but at maximum speed (a little over 80 m.p.h.) the engine is running at approximately 5,000 r.p.m. and developing rather less than 55 h.p. It can be calculated from this that to achieve 100 m.p.h. the engine would have to run at 6,300 r.p.m. and develop about 95 h.p. An increase of about 25% in engine r.p.m. seemed a rather high figure at which to aim, although the piston speed would remain quite modest at around 3,000 ft./min. Nevertheless, certain opportunities of adjusting the overall engine-to-road-speed relationship presented themselves, and as a first stage towards the final target effort was concentrated on the cylinder head and inlet and exhaust manifolding.

Improved Breathing

There are three inlet ports on the Zephyr and it seemed sensible to attack these most directly by attaching three separate carburetters to them. Slightly inclined $1\frac{1}{4}$-in. S.U.s were chosen and after the car had been consigned complete to the Castelnau works of Boon and Porter they machined three short stub pipes with a tapering internal bore to which the carburetters were attached. Whilst these pipes were being made the combustion chambers and inlet and exhaust ports were smoothed out and the face of the cylinder head cut back by 0.040 in. This last step increases the compression ratio to rather over 7.5 : 1 and is about the limit which can be reached bearing in mind the offset design of the combustion chamber.

Tested in the intermediate gears, the standard engine was found to be capable of 5,000 r.p.m., but in view of the extra engine revs which were being sought it was thought desirable to increase the valve spring strengths and this was achieved by fitting double springs to each valve, the experience in this matter of Allards with their Palm Beach models being drawn upon.

Having thus assured improved breathing on the inlet side of the engine, the next question was that of giving a free exit to the exhaust—a matter in which I have been particularly interested since my early Bugatti experiences.

The standard Ford exhaust system is a most ingenious piece of production engineering for there are six recesses formed in the cylinder head where the six exhaust ports emerge, and into these recesses is clamped a horizontal pipe, suitably perforated to receive the gas.

With no basic information to go upon I was somewhat sceptical as to the ability of this lay-out to meet a substantial increase in engine power without finding itself in difficulty from a volumetric, or thermal, viewpoint or perhaps both. In order to remove this hazard I decided to replace the standard lay-out by something which would guarantee free gas exit at any possible power output.

On the suggestion of Mr. " Bob " Porter (who played a very active part in developing the lay-out) some cast-iron blocks were made which were machined to fit the existing recesses, being locked to the block by Allen set screws fitting into the threads normally used to hold the exhaust pipe clamps. These blocks are pierced and form in effect an added part of the head which is an extension of the exhaust ports. To them are bolted entirely separate exhaust manifolds for the front three and back three cylinders. These separate manifolds discharge into separate tail pipes, each of which is fitted with a Servais silencer.

Giving a Free Hand

Having thus attended to the exhaust system there remained the question of air filtration and silencing and here the Vokes Co. asked if they could have a free hand to try out some ideas they had in this connection. This suggestion having been accepted they fitted a very large air cleaner in which the inlet faces backwards. This inspires warm air (which is probably no disadvantage taking the whole year round) and enables the whole filter element quickly to be withdrawn from the nose of the car. The filter has a central offtake and from it radiate three very handsome polished aluminium pipes leading to the three S.U. carburetters.

The revision of the manifolding would have been difficult without the removal of the accumulator from the nearside of the bonnet. The opportunity was therefore taken of replacing the standard battery with a narrow 12-volt Lucas type mounted immediately behind the rear seat and thus directly above the rear axle, which finds the additional 60 lb. load a help in reducing wheelspin.

Although not bearing directly upon the performance of the car certain other changes were also made in the electrical system. An S.U. pressure-type pump was fitted in the luggage boot next door to the battery and this supplies fuel by an entirely separate piping up to a changeover tap beneath the bonnet. This tap also communicates with the standard mechanical fuel system so that there are up to this point two entirely independent fuel lines, each with its own pump. A short transverse pipe feeds the three carburetters through a large Vokes petrol filter and Lodge HLNP plugs are fitted.

The car came back onto the road with all these modifications on May 9 and in the ensuing two months covered about 3,000 miles, and it is a great tribute to the thoroughness with which the work was carried out by all the " contracting parties " that the only mishaps have been one broken petrol pipe and the working loose of the float chamber on No. 1 carburetter.

It cannot be said that the improvements in performance derived from these changes have come up to expectations, for they have in fact been exceeded by a comfortable margin.

107 on the Clock

In brief, the maximum top-gear Tapley figure has been raised from 270 lb./ton to 290 lb./ton which is the equivalent of increasing the maximum top-gear gradient from 1 in 8.4 to 1 in 7.8. Correspondingly, in the upper part of the speed range the pace which can be maintained up a steady gradient of 1 in 20 has been raised from 69 to 80 m.p.h. That is to say, the car will now climb a gradient of this order at approximately its previous maximum speed.

The sheer flat-out maximum is now limited by the rather low effective gear ratio but between 106 and 107 m.p.h. can be shown on the speedometer, representing a true maximum speed of a little over 95 m.p.h., or a fraction under 6,000 r.p.m. Mentally, the thought of a low-price, mass-production engine rotating at this speed is a little frightening; physically, if one did not know the crankshaft speed one could never guess it, as the engine is very smooth, as it is indeed throughout the whole range. This notwithstanding, the great temptation to run for long periods at so high a speed has been resisted, and pending the arrival of new wheels and tyres of greater effective radius the most marked benefit in performance has been found in the rapid acceleration between 60 and 85 m.p.h. and the ability of the car to cruise very comfortably at rather over its standard maximum. This in turn gives hourly distances with relaxed and comfortable driving which were hitherto only possible with considerable mental and physical effort, and the improved performance at the top end adds materially not only to the ease but also to the safety of the day's motoring.

The previous carefully recorded fuel consumption has been impaired by about 15% but it is hoped that careful tuning in conjunction with the forthcoming higher effective " gear " will improve this factor.

It was found that sustained high engine speeds resulted in some loss of oil pressure both on the straight and when rounding corners, but this somewhat alarming experience has been eliminated, for all practical purposes, by the use of Castrol G.P. oil. Spring damping has been increased by a resetting of the front dampers and the use of larger rear dampers with a higher setting, this work being carried out in collaboration with the original shock absorber manufacturers, Armstrong Patents, Ltd. To sum up, there is some need for additional development work, and it is hoped, for example, that some changes in the braking system will give a greater margin against fade than is now present. Taken as it stands, however, the car is a remarkable proof of how much can be gained by applying classic methods of modification to a basically sound design, and although going up from 95 m.p.h. to 100 m.p.h. will require 17½% more h.p., a Zephyr Century appears to be within the bounds of possibility.

SPEEDING the ZEPHYR

A £75 " Raymond Mays Conversion" to give over-90 m.p.h. performance from the Ford Zephyr Six

READY FOR FITTING to the Ford Zephyr Six engine is the modified cylinder head with triple carburetters, suitable controls, short inlet pipes with balance inter-connections, and additional inner valve springs. The right-hand picture shows that with the battery removed from alongside the engine to a box in the rear locker, there is ample space for access to the carburetters.

CONSIDERABLE interest has been aroused by our recent description of how the Ford Zephyr Six saloon run by a member of our staff has been modified to give much-increased performance. We have now had the opportunity to make some tests on another triple-carburetter Ford Zephyr Six, incorporating slightly less elaborate tuning modifications which are being offered at a fixed price of £75 by Raymond Mays and Partners, Ltd., of Bourne, Lincs.

As on the car already described, the basis of the Raymond Mays conversion is replacement of the single Zenith downdraught carburetter by a trio of horizontal S.U. type H4 carburetters. To take advantage of these carburetters, however, a substantial amount of work is done on the Ford cylinder head casting, the ports being opened out to an improved shape and polished, the combustion chambers polished, and the head face machined to raise the compression ratio from 6.8/1 to 7.6/1. All the normal Ford valve-gear parts are retained, but additional springs inside the main valve springs raise the r.p.m. which can be used without valve-bounce.

Top Gear Flexibility

No change from the simple Ford exhaust manifold is made in this conversion, but a Servais silencer of the straight-through type reduces exhaust back pressure. To allow room for the front carburetter, the 12-volt battery is removed from its position alongside the engine to a box inside the spacious luggage locker, a removal which also gives more favourable weight distribution for most driving.

On the roads of Lincolnshire we were able to appreciate how much this conversion increases the already high performance of the Zephyr Six, and does so without detracting seriously from its essentially " touring car " characteristics. Top gear flexibility at speeds of the order of 10 m.p.h. is quite unimpaired, and although the absence of carburetter air intake silencers allows a certain amount of healthy power roar to be heard during acceleration the car is certainly not objectionably noisy.

Facilities for making precise maximum speed tests on this car were unfortunately not available, but extrapolating from speedometer calibrations over the 20-80 m.p.h. range of speeds it was clear that speeds in the region of 95 m.p.h. were being attained during hard driving on open roads. As the acceleration figures published on this page show, the 80 m.p.h. pace which is approximately the maximum of a normal Zephyr Six was attained from a standing start in under 30 seconds by this " Conversion " model. This car was stated to be running on Regent T.T. premium-grade pump fuel, and showed no tendency whatever to pink.

FORD ZEPHYR SIX

Top Gear Acceleration	*The Motor* Road Test 6C/51	Raymond Mays Conversion
m.p.h.	secs.	secs.
20-40	8.2	7.1
30-50	9.4	7.8
40-60	11.8	8.7
50-70	16.0	9.9
60-80	—	13.0

Acceleration through the gears.	*The Motor* Road Test 6C/51	Raymond Mays Conversion.
m.p.h.	secs.	secs.
0-40	8.8	7.7
0-50	14.0	10.2
0-60	20.2	16.3
0-70	29.9	22.1
0-80	—	29.3

The adequacy of Ford Zephyr Six road-holding, aided only by the slight rearward transfer of weight, is most praiseworthy at increased cruising speeds, the car being comfortable and eminently controllable although there is perhaps a feeling that too much should not be asked of the brakes. The car tested, although well maintained as befits the personal car of Raymond Mays, was stated to have already covered 30,000 fast miles without showing signs of deterioration, and whilst there was some propeller shaft vibration at medium speeds the short-stroke engine gave no sign of objecting to sustained high r.p.m. We were not able to check fuel consumption ourselves, but we are assured that at any given speed the conversion gives normal or better - than - normal fuel economy: when cruised fast on French roads to average just over 60 m.p.h. from Rheims to Dijon the fuel consumption was stated to be 20.8 m.p.g.

Hardly ranking as a sports car with no more than 50 m.p.h. usefully available in the highest indirect gear, a Ford Zephyr Six with this conversion does nevertheless provide quite remarkable performance for less than £830, inclusive of British purchase tax. Hitherto, the various cars which have already been subjected to the Raymond Mays conversion have been dealt with at the works at Bourne, and the inclusive price of £75 is based upon this arrangement. Plans are however being made for the supply of kits of parts on an " exchange " basis, so that distant customers may have their cars converted rapidly by any competent garage.

The three S.U. carburettors, with individual air cleaners, fitted to a Ford Zephyr six-cylinder engine. The standard exhaust manifold is employed.

FIVE STAR PLUS

IMPROVED PERFORMANCE FROM A FORD ZEPHYR WITH A THREE - CARBURETTOR CONVERSION

IT is well known that the Ford Zephyr has more than adequate performance for a standard saloon car of medium size. It can be as fast from point to point as a number of cars in a higher price range and at the same time gives an economical fuel consumption figure.

With present-day traffic conditions and their attendant problems it has been evident for some time that rapid acceleration is becoming more important than overall maximum speed. For owners who wish to improve the already brisk acceleration of the Zephyr, the Ace Service Station (London), Ltd. is marketing a three-carburettor conversion set for this car. The set consists of three S.U. Type D2 downdraught carburettors, with curved cast aluminium alloy inlet pipes, which are fitted with interconnecting balance pipes. It is not necessary to do any drilling or tapping and it is reasonably claimed that the conversion set can be fitted using only the standard tool kit supplied with the car.

The price of this conversion set complete with the necessary fuel pipes, throttle linkage and gaskets is £39 10s, while for a further £3 an air cleaner can be supplied for each carburettor. It is not necessary to reposition the battery, as the front carburettor clears this unit by approximately 1½in.

A 1953 model Zephyr fitted with this modification was recently tested by *The Autocar* and some interesting performance and consumption figures were obtained. In addition to the basic conversion described, double valve springs and a Servais silencer had been fitted to the car tested, and 0.040in machined off the cylinder head. This work can be carried out for a further £6 15s 6d, and has the effect of increasing the compression ratio from 6.8 to 7.4 to 1. On premium grade fuel, pinking was not evident, even when accelerating hard in top gear.

Acceleration through the gears showed a marked improvement, as would be expected, over that of the standard Zephyr saloon, while the top gear maximum increased by the order of 5 m.p.h. A certain amount of roar was noticeable when accelerating hard, resulting from the intake silencer being replaced by small circular air cleaners, but instead of being objectionable it gave the driver a rather satisfying indication of the performance. It was quite possible to spin the rear wheels in starting on dry surfaces if the throttle was opened suddenly, and, with this in mind, due caution was exercised on wet roads. Overtaking and the climbing of main road hills at what would be a creditable speed on the flat was quite an experience.

The car tested took 8.1 sec to accelerate from 20 to 40 m.p.h. on top gear, while the standard Zephyr tested by this journal in 1951 took 9.2 sec. From a steady 30 to 50 m.p.h. in top gear 8.5 sec were taken, as against 10.2 sec.

From rest through the gears to 40 m.p.h. took 8.5 sec and 50 m.p.h. was reached in 12.6 sec, while 70 m.p.h. in 24.9 sec showed an improvement of 7.2 sec over the figure for the standard model.

If advantage is taken at all times of the increased performance, fuel consumption is bound to suffer. Driving the car hard at all times, including the test work and journeys across London, produced a consumption figure of 16.3 m.p.g. On the other hand, on a cross-country journey of about 100 miles, cruising at a genuine 60-65 m.p.h. and using the acceleration to advantage when required, this figure improved to 20.7 m.p.g., while normal weekend motoring round country roads gave 22.8 m.p.g.

This last figure tends to bear out the supplier's statement that consumption will not be any greater than with one carburettor, provided that maximum use is not made of the increased power all the time. There is no doubt that this fitting does give the Zephyr an added sparkle at a comparatively low price, the complete conversion, as on the car tested, costing £55. Full details can be obtained from Ace Service Station (London), Ltd., North Circular Road, Stonebridge Park, London, N.W.10.

A WHEELS ROAD TEST OF THE

FORD CONSUL

This moderate-sized family saloon gave us a real surprise. It exceeded 80 m.p.h.—5,400 r.p.m.!

By BARRIE LOUDEN

OUR major impression of the Ford Consul was its amazing top speed. On test it exceeded 75 mp.h. comfortably and on one of the time runs bettered 80 m.p.h., reaching an engine speed of 5,400 r.p.m.—an outstanding performance for a five-passenger English car of 1½ litres.

It is worth noting that there had been no maintenance on this car for 300 miles before the "Wheels" test.

The Consul we tested, incorporates many minor modifications made since the car was introduced, has an outstanding performance, good road manners, steering which cannot be faulted, a large boot, and bench seats front and rear. It is a very good small family car.

From the driving seat the driver has excellent vision all round and one immediately feels at home behind the wheel. Although rain was encountered during the first part of the test we found adhesion was well above average and allowed high speeds to be maintained regardless.

The unique, long coil-spring front suspension gives a firm ride but a comfortable one. It absorbs rough surfaces very well and noise is very low. Pothole areas were negotiated without discomfort to the passengers or objections from the vehicle.

The car handles very well indeed at speed, and can be taken through corners in sports style. Body roll is present, but not to a degree which calls for criticism. Tyre squeal is evident and is too readily provoked. Close attention to tyre pressures is a must for Consul drivers.

Under pressure the rear wheels will break away slightly. This is easily corrected, particularly with help from the lively engine.

Position of the steering wheel is excellent, as is the steering itself. It is both light and accurate, with only 2½ turns of the wheel from lock to lock. The car does not noticeably under or over-steer.

Rake of the wheel is more upright than usual, and gives leg-room around the steering column gear lever for a tall person.

The over-square engine is surprisingly lively and flexible. It is quiet at all times, free from vibration and carburetter hiss. Pulling power in top gear is equal to many larger cars.

Overheating—a bugbear for many cars in Sydney at least when the temperature was 107 — did not occur even after the speed runs on a scorching day.

Engine accessibility is excellent. The battery, often tucked away in an inaccessible position, is mounted forward near the radiator.

Maximum speed shows some slight roughness, which we were inclined to put down as wheel misbalance. On the other hand it might have been the substantial margin of revs the engine was pulling above its peak figure.

The gear change has the shortest travel of any steering-column lever we have tried. The synchromesh cannot be over-ridden and the driver does not have to lean forward to select gears.

The ratios were well-balanced, and the box itself is very silent in operation. The clutch is connected to the gearbox by a hydraulic linkage, and while having a short throw calls for a high pedal pressure.

The brakes are very good, and although pedal pressures are heavier than some cars the action is smooth and positive. There was no sign of fade during the test. Panic stops from about 78 m.p.h. were accomplished in a straight line without wheel lock. A pleasant surprise was that the car did not stand on its nose during heavy braking.

(*Continued on page* 73)

TECHNICAL DETAILS

ENGINE :
4 cyl., o.h.v., bore 79 mm. x 76 mm. stroke, capacity 1,508 c.c. Comp. ratio 6.8:1. 47 b.h.p. at 4,400 r.p.m. Speed at 2,500 ft./min. piston speed, 75.5 m.p.h. Max. torque 74 lbs./ft. at 2,000 r.p.m. 15.1 m.p.h. per 1,000 r.p.m. in top gear. Single downdraught carburetter, Zenith. AC mechanical fuel pump. Cooling by pump and fan. Radiator 16⅜ pints, sump 6½ pints.

TRANSMISSION :
Single dry plate clutch, hydraulic pedal linkage, three-speed gearbox, synchromesh on top tow ratios. Ratios 4.556, 7.704, 14.898. Steering column gear lever. Hypoid bevel final drive, open propeller shaft.

SUSPENSION :
Independent in front by coils, integral with hydraulic telescopic shock asborbers. Semi-elliptic rear springs with hydraulic shock absorbers and anti-roll bar.

BRAKING :
Girling hydraulic, 2-leading-shoe in front, 9" drums, friction lining area 121 sq. in. Pull-out parking brake mechanically linked to rear wheels.

CHASSIS :
Combined body-and-chassis pressed steel unit.

IGNITION :
Lucas 12 volt, 57-amp-hour battery.

PETROL TANK :
9 gallons.

WHEELS :
5-stud, disc, 13", Dunlop 5.90 tyres.

STEERING :
Worm and Peg, 2½ turns from lock-to-lock, turning circle 40 ft.

MEASUREMENTS :
Wheelbase, 8 ft. 4 in.; front track, 4 ft. 2 in.; rear track, 4 ft. 1 in.; height, 5 ft. 1 in.; length, 13 ft.; width, 5 ft. 4 in.; ground clearance, 6½ in.; unladen weight, 21¾ cwt.

PERFORMANCE :
Acceleration from rest through gears:
From rest to 30 m.p.h., 5.25 sec.; rest to 40, 10 sec.; rest to 50, 16.75 sec.; rest to 60, 25 sec.
Acceleration in top gear:
From 10 to 30 m.p.h., 11.5 sec., 20 to 40, 11 sec.; 30 to 50, 10.5 sec.; 40 to 60, 14 sec.
Standing quarter mile:
23¼ sec.
Maximum speeds:
In 1st, 20 m.p.h.; 2nd, 50 m.p.h.; top, mean of three runs, 77 m.p.h.; best time 81 m.p.h. (5,350 r.p.m.).
Fuel Consumption:
32 m.p.g. on test.
Laden weight as tested:
24 cwt.
Speedometer correction:
Nil.

CONSULAR

COMMENT O

Clean lines of the current Ford Consul combine a smart appearance with the practical attribute of a body that is easy to clean. The flattened wheel arches conform to the horizontal styling.

DIFFERENT examples of the same model of car made by almost any manufacturer have slight differences despite the increase in uniformity that results from modern manufacturing methods, and on the present occasion the car is not just "a Consul," but RMV 162, an opal blue example that was first registered at the end of 1952 and came into my hands about a year ago. As many miles are covered for various purposes in other cars, this Consul has so far journeyed to the relatively modest total mileage of a little more than 23,000, but this represents enough work done for some conclusions to be drawn.

Before commenting on the car's good and bad points, it must be admitted that much interest has been provided by the slight modifications that have been made from time to time. Reliability has always been of particular importance and cost has been considered before any modification has been made (having in mind both first cost and future effect on m.p.g.). No really costly modifications have been made, yet the table of performance data shows a considerable improvement step by step despite the raising of the overall gear ratios. Apart from modification for improving the performance, a number of additions has been made to the standard equipment.

Family Favourite

The Consul as a model is now thoroughly familiar on British roads, and I share the widespread opinion that it is a particularly good looking car. It has a very smooth shape that is in all ways really practical; it looks businesslike, and it is provided in a usefully wide range of pleasant colours. And in addition to first impressions being good, the Consul is very much a car that grows on its owner for a number of sound reasons. It really will seat six average size people without undue discomfort; it has an excellent luggage locker of good shape as well as size; it is easy and pleasant to drive and performs very well; and its good shape is easy to clean. Criticisms are not numerous, but it must be recorded that the visibility past the windscreen pillars is not good, the steering is a little woolly, the weight distribution is not ideal, and the layout of the minor controls could be improved. On RMV 162 there is also an unpleasant vibration in one certain speed range.

I remember that on taking the car over I at once liked the smoothness of the "square" engine and the general liveliness. The fact of the gear box having only three forward speeds allows the steering column gear change to have only two "layers" and this helps to provide one of the best gear changes of the steering column type currently in use. But first impressions also included dissatisfaction with rear wheel adhesion. Light weight at the back coupled with small (13in) wheels combined to make wheelspin occur too easily, and wheelspin was particularly easily contracted in roundabouts and similar fairly sharp bends of the type that can best be taken in middle gear, and that meant tail wag. Naturally this lack of adhesion was most noticeable in wet weather and on already slippery surfaces. Another early impression was made by the horn, which kept the car within the law but was of little use for overtaking Continental *camions*.

It is but fair to the car to give some indication of the type of use with which it has had to contend. It has been necessary on a few occasions to drive very hard indeed for anything from 200 to 400 or more miles, virtually at a stretch to keep within the schedule of rally coverage, or to catch the appropriate boat after a foreign race meeting. Narrow lanes and tracks have been tackled repeatedly in following the course of trials and night rallies, and by no means has the car always had the advantage of being lightly loaded. Yet, apart from the axle, of which more presently, nothing has "broken" or even come loose, and the paintwork has not faded.

Polish is hardly ever used on the bodywork, for a wash and leather down restore the sheen admirably. Unfortunately the same cannot be said for the chromium, which is kept in good condition only by attention at never less than weekly intervals. When the car has been deserted for more

In this engine view the air cleaner has been removed to expose the four-branch exhaust manifold. The stubs are not very sharply raked, because of the proximity of the bulkhead to the rear pipe, but the manifold makes a considerable difference to the performance by permitting better breathing.

SERVICE

By MICHAEL CLAYTON

TWELVE MONTHS' MOTORING IN A SLIGHTLY MODIFIED FORD CONSUL

RMV 162 has acquired two extra lamps and a bonnet *motif* without, it is hoped, making the appearance too fussy. The lamps perform a necessary function with considerable success.

Ordinary side lamps have been let in to the body at the rear, where they act efficiently as reversing lamps without being vulnerable.

than a week in damp weather in favour of the temporary use of another mount it has been necessary for its chromium to be cleaned, and on a number of occasions brief halts on a rally have been used to keep rust in check. With this care the chromium is not pitted to any appreciable extent, but one has the feeling that it is waiting only for the driver to turn his back for a month or so, whereupon it would get into a condition from which there is no complete return.

Mechanical changes began when the crown wheel and pinion in the back axle became suddenly noisy. The fault was a bad oil seal where the shaft enters the axle casing, but the manufacturers identified RMV as one car in a small batch in which defective seals had inadvertently been used, and made good the damage. As the crown wheel and pinion had to be replaced, however, the decision was made to substitute the Zephyr components, giving a top gear ratio of 4.444 to 1 instead of RMV's 4.625 to 1. The current Consul has a ratio of 4.556 and a change to the Zephyr ratio would make an almost imperceptible difference, but on earlier Consuls the difference, although small, is worth having. Before power output was increased the acceleration inevitably suffered from the new ratio, but this was a state of affairs that did not last long.

Means to an End

Last year the air was thick with conversions provided by various firms to improve the already creditable Consul and Zephyr performance, and two carburettors, with perhaps some work on the cylinder head and stronger valve springs, provided a popular basis. But the exhaust manifold on the Consul is as unusual as it is ingenious, and I was intrigued by an alternative to the standard layout as a starting point for tuning. As has been said already, reliability and m.p.g. have been considered specially important, and appearance suggested that the standard Ford exhaust manifold could be replaced by something that would cause less back pressure and thereby make for a more efficient engine. A relatively inexpensive four-branch manifold manufactured by the appropriately named firm called Consul Productions was fitted, and the results were very satisfactory, as can be seen from the data table.

There was one snag which has since been overcome. Larger main and compensating jets were recommended and used, to keep the mixture properly balanced, but it has since been found that these are not really necessary except for a change of main jet on the earlier Consuls. On the currently produced cars the main jet is larger than that fitted originally and can be used satisfactorily with the manifold. And further experience with the manifold has shown that it is not really necessary to increase the size of the compensating jet. RMV has always started particularly easily and, running on

Through gears	Standard Consul	With Zephyr axle ratio	Zephyr ratio and exhaust conversion	Zephyr ratio, exhaust conversion and increased compression
m.p.h.	sec.	sec.	sec.	sec.
0–30	7.7	8.2	7.3	5.9
0–50	19.7	23.8	17.6	15.2
Top gear				
m.p.h.				
10–30	12.7	12.4	10.2	10.0
20–40	13.1	15.3	10.0	9.9
30–50	15.1	13.2	11.2	11.1
m.p.g.				
quiet driving	28	30	27	29
hard driving	24	25	23	23

the flat curve Duckham's Q5500 engine oil (which, to put it another way, does not "thicken up" when it is cold), it has been capable of giving almost full power as soon as it is running. I find that with a standard size compensating jet in conjunction with the four-branch manifold the engine still fires at once but needs a mile or so on the road to warm up before it will pick up on acceleration without any trace of hesitation. This is of no consequence as, in any case, the engine should be permitted to warm up before full power is used.

Despite the use of the Zephyr axle ratio the car now performed better in acceleration tests than it had originally, but there was still one more thing that was clearly worth doing. In its standard form RMV 162 would run quite well on ordinary grade fuel, even though it was a little sweeter on first grade, and to take more advantage of the high octane fuels now available it was decided to raise the compression ratio. In one or two twin carburettor conversions was included the machining of more than 0.060in (60 thou.) off the face of the cylinder head, and I know of one case in which 0.070in has been taken off and the head polished without any ill effects being immediately obvious. However, it was decided to be less ambitious, for the aim was to increase the efficiency of the unit rather than adopt a sports car engine tuning programme that might lead to trouble on the kind of work for which the car was required.

The compression ratio was finally raised from 6.8 to 1 to 7.5 to 1 by the removal of 0.054in. It is interesting to note here that the removal of this amount of metal provides, in effect, a cylinder head with the same compression ratio as that catalogued by the Ford company and intended by them for use in high-altitude countries. The result of this modification can be seen in the table. Top gear acceleration does not seem to be much improved, but this is perhaps because conditions were particularly good for the earlier set of figures, whereas it was rather windy when the final figures were taken. In indifferent conditions the top gear acceleration readings would be affected most adversely. In its

CONSULAR SERVICE

The original tyres were replaced by the Michelin Super Comfort type seen on the left, but the latest Michelin for the Consul and Zephyr is shown on the right. It incorporates the same tread pattern as that used in the X tyre (which has a steel inner mesh), and is intended to be silent without loss of grip.

current form the car is not perhaps *quite* so flexible as it was originally, but the engine is still sweet running even at idling speeds and there is no difficulty whatever in starting when cold or hot.

A comparison is interesting between the present performance and that of the standard Zephyr and of Consuls that have been subjected to more expensive conversions involving the use of two carburettors. Most of the present figures are only about 1 sec higher than those obtained on the Zephyr submitted to *The Autocar* for a full Road Test, and the 0-30 m.p.h. figure is only 0.4 sec higher. More comprehensive tuning conversions on the Consul provide acceleration figures up to 50 m.p.h. that lie between those who RMV and for a standard Zephyr. Of course, it is reasonable to suppose that where much work is carried out on the cylinder head porting the performance at higher speeds would be better.

As with so many other cars, m.p.g. depends very much on the way the car is driven. With a fair amount of town driving in which good use is made of the performance, and with high speed used on the open road, the m.p.g. drops to the poor figure of about 23; while cruising at about 55 to 60 m.p.h. on the open road, and without too much fierce acceleration, the figure rises to a satisfactory 29. The most important influence on m.p.g. is certainly acceleration, and to achieve an m.p.g. nearer 30 than 20 even with the use of a high cruising speed on the open road it is enough to avoid overtaking fiercely when it is not necessary and to make a smooth progress through towns.

The Consul's brakes are very powerful, but on this car there was a tendency on wet roads, and under heavy braking, for the front to slide off its line. This, and the wheelspin already referred to, prompted the thought that tyres designed with the emphasis on grip would be worth while. Some

The arrangement of the minor controls is not very neat or compact, and on the car here concerned is further complicated by the knobs for the auxiliary lamps, one of which (black) is seen on the side of the cowling round the steering column, and the control for the windscreen washer (extreme right). An ash-tray is attached to the swivelling window.

experiments had already been made with ballast in the luggage locker, but they were not altogether satisfactory. The practice of making the weight distribution more even by adding weight at the rear would probably be more worth while for Zephyr owners, and on either model it is easily accomplished with coal weights. These, weighing half a hundredweight each, are little more than twice the size of a brick, and have a flush fitting handle in the upper surface. They fit into the rear corners of the locker very conveniently. With the rear of the Consul weighted in this way, tail wag as a result of wheelspin was reduced slightly, but the very idea of adding weight is unpalatable and it was decided that a different tyre tread would be more satisfactory.

Tyre wear with standard tyres was very creditable. Despite hard driving there was still tread on the tyres at over 19,000 miles. Wear was even because the tyres had been changed round at intervals, but it was noticed that those at the rear wore a little more quickly than those at the front, which may be taken as a compliment to the i.f.s. or a criticism of the wheelspin. At 19,000 miles the treads were thin, and although in summertime they would have been good for at least another 1,000 miles a winter change was made for safety's sake, and this time Michelin Super Comfort were fitted.

Grip

These are low pressure tyres with a frilly tread that takes a very strong grip on the road. Instead of the 28 lb per sq in all round that was used previously, pressures of 22 lb and 24 lb were recommended for the front and rear respectively. The front tyre recommendation was 2 lb lower to ensure maintenance of understeer characteristics. The grip with these tyres is certainly very good, because there are so many frilly little edges to contact the road, but for this very reason the life is expected to be a little shorter than with some other tyres. However, for hard driving, grip is particularly important. The handling of the car has changed inasmuch as one has to "take up the slack" in the tyres while lining up the car for a bend, but, this having been done, the stability is very satisfying. The main criticism of the Michelins is the increase in noise. This, a low-pitched growl, is not noticed so much at high speed, other than on smooth, wet surfaces, as on town roads, and particularly during such manœuvres as coasting up to a red traffic light.

At the recommended low pressures the rolling resistance of the car is virtually unchanged because the tyres are so supple, and any slight rattles in the bodywork disappear. But as extra tyre wear will not take place at slightly higher pressures I like a compromise of 25 lb and 26 lb, losing some of the extra comfort but keeping the grip and avoiding a little of the need for "taking up the slack" on bends. These pressures being only 1 lb lower at the front, there is but slight understeer; as a result the steering is more responsive and the car corners very well.

Stop Press: Coincident with the writing of these comments comes news of a new Michelin tyre for Zephyrs and Consuls that is superseding the Super Comfort, and a set has been fitted to RMV experimentally. These new tyres have the same tread pattern as the Michelin X (which has steel mesh under the tread), with small but deep cuts spread more evenly in the tread than they are in the Super Comfort. The manufacturers claim the elimination of the growling noise without any reduction in grip. Experience of them on RMV is too brief to justify attempting to give a concrete opinion of them, but certainly it may be said already that the growling noise has gone. To test the grip I must wait for rain.

At a mileage of over 23,000 it should be possible to judge to some extent what the Consul is like for the man who wants a car of this size at moderate cost and as a long-term investment. RMV is still quite young, in both age and mileage, but this is about the stage at which a car ceases to be "new." It is the stage at which before the war many people parted with their cars in favour of new ones, considering this method to be the most economical and satisfactory for reliable service with hard use. The case history of this car is creditable. True, the axle went early in life, but this might reasonably be called a freak fault that was put right without charge by a local agent. New tyres were needed in the neighbourhood of the 20,000-mile mark, and with quiet driving the tyre life should be very good indeed. Nothing else has been done to the car other than good and regular routine servicing. The original battery, brake linings, radiator, steering box, clutch, and so forth, are still giving entirely effective service. There are one or two very slight body rattles that are about to receive attention, but the only real fault about to be tackled is the spring dampers, which are ready for reconditioning.

There is every indication that the engine will chalk up a really high mileage before major overhaul. At 23,000 the oil consumption is virtually negligible. A recent run of about 1,200 miles was started with the sump inadvertently slightly overfilled, and was completed without the addition of fresh oil, with the level only slightly below the full mark. Even under sustained fast driving it is doubtful whether the engine consumes more than a pint in 1,000 miles.

The heater-demister and radio installations are very neat, and the units themselves are commendably effective. The parcel shelf is good, but some controls are scattered; the starter knob, for example, is on the left of the heater control box, virtually on the passenger side of the car.

In only one respect is RMV 162 disappointing, and investigation has shown that the fault is not universal among the other examples of the model. This is an appreciable vibration that comes in with the present gearing at about 60 m.p.h. (or 63 on the speedometer, the earlier and greater optimism of the reading having been reduced by the change in axle ratio). This persists up to about a true 66 and is particularly annoying because this is just about the speed range at which the Consul will normally travel happily on a long journey. In the high fifties the car is but loafing along, but it takes time to get through the period of vibration and in England it does not often seem worth while doing so because speeds near a true 70 cannot be maintained for very long, nor are they desirable from the viewpoint of the machinery. The vibration has the symptoms of out of balance in the propeller-shaft, but the balance of the shaft and the mountings of the engine and gear box have been carefully checked and found to be correct. Investigation continues. . . .

The inadequacy of the original horn started a chain of minor changes in the equipment. Audible warning of approach is now effected by a pair of Lucas Continental horns. These are of the Windtone type and are penetrating. The best mounting for them would be behind the neat radiator grille, but this area is none too accessible, and the

Horns of increased power are installed under the bonnet, and the windscreen washing container and jets can also be seen. The heater unit, an optional extra, is alongside the right valance. Engine accessibility is good by current standards.

horns are, therefore, attached to the underside of the bonnet, in the position of the original single horn—the necessary extra hole was already drilled, it was found! They are effective for most purposes and have the added advantage that the type of driver, perhaps of another Consul, that objects to being passed by a similarly powered car is sometimes momentarily deceived by the rather expensive sound from behind.

After the horn came auxiliary lamps, the most important being a Lucas "Flamethrower" driving lamp and a Marchal 630 wide-angle lamp intended for fog and for driving along twisting roads. The Lucas lamp was mounted on the right side so that it would shine as near the centre line of the road as possible with its very narrow beam, and the Marchal was mounted on the left. The "Flamethrower" is more useful even than it was at first expected to be. Its use is often thought to be confined to high speed night travel on fast roads, but I find that it is particularly useful on those typical English smaller main roads that are well surfaced but which contain many sharp and unexpected corners. The Marchal is ingenious. The beam is very flat directly in front of the car, but it curves up at the sides. It does not cause excessive back glare in fog, but it lights up the verges and hedges for some distance ahead. It shows you where you are as well as where you are going, and it does not seem to dazzle oncoming drivers. And on Alpine roads it lights the way "round" the hairpins.

Side Lamps at the Rear

For the solution of the reversing light problem I am indebted to a suggestion made by the Consul-owning producer of the exhaust manifold. Instead of the more usual small separate lamp, two ordinary side lamps are fitted one on each side of the number plate. They are not vulnerable and they provide an even spread of light to the rear. Also, they are not expensive. Over-riders were also fitted at the rear in the hope of warding off traffic that may be caught out in emergency by the power of the Consul's brakes.

An early acquisition was a windscreen squirt apparatus. This has always seemed to me to be most necessary on any car, and has been used very frequently in the winter and also for clearing the screen on the move when driving into a setting summer sun. A Wilmot-Breeden *motif* has appeared on the bonnet. Its original purpose was to act as a mounting for a number of insect deflectors that were tested, but it remains as a pleasant piece of ornamentation. The driver's side swivelling ventilator window has grown an ash-tray. In the Consul both gear lever and hand brake are on the left, which means that if one smokes cigarettes at the wheel the right hand is the one to use. The only other items that come under the heading of additional equipment are the optionally extra heater-demister and an Ekco radio. Both earn high praise. The demister really demists, the heater really heats, and the radio is powerful, has good tone, and is particularly free from interference.

RMV 162 should have a long and hard, but efficient, life ahead.

FAST CRUISING.—". . . the Ford sped easily down the long straights between rows of poplars . . . with the speedometer needle gently undulating between 70-80 m.p.h. . . ."

FIVE DAYS' HARD

. . . but not for the driver, on a 1,700-mile Continental journey

By E. H. Row

"LET us not," said George, "stay the night in Paris"—a less silly suggestion than at first appeared, for so to do without (*a*) spending more than one can afford, or (*b*) getting to bed later than one should, is, to my mind, "agin' nature," and it was necessary that we did neither of these things. Our assignment was to report on the Paris start of the Tulip Rally, proceed to Belfort to pick up the competitors as they began the southern loop of their run, drive with them and observe their doings for the ensuing 24 hours; next day proceed to Geneva to deposit photographer George Moore's films on an aeroplane for England and himself on a train for Italy and the Mille Miglia; and then catch the night ferry from Dunkirk the following evening, finding time in between to compose a story on what one had seen. Such programme called for conservation of both francs and energies during the early stages and thus it was that we found ourselves taking the evening air on Chantilly's racecourse instead of being taken for a financial ride in some Parisian night haunt.

Transport for the occasion was a Ford Zephyr convertible, offered by the manufacturers and accepted with alacrity for two reasons. First, *The Motor* tests have shown the Zephyr to be an eminently suitable car for fast, hard motoring, and secondly because the model offered *was* a convertible and thus, to my mind, enabled the very best to be derived from motoring, no matter what the weather conditions.

The necessity for seeing the Paris starters off on their night run to Nurburgring resulted in our not leaving the city on our way to Belfort, where we were to pick them up again, until four o'clock in the afternoon. Thus, when the bonnet of the Zephyr was eventually pointed down N.19 toward Troyes and the south-east, the first 20 miles or so were fairly littered with Simcas, Renaults and Citroens of various types and ages, bringing hundreds of exuberant Parisians back home after a day in the country—all very individualistic and exciting. Once clear of this tangle and given its head, the Ford sped easily down the long straights between rows of poplars, glowing gold in the evening sunlight, with the speedometer needle gently undulating between 70 and 80 m.p.h. and one's accelerator foot on the rest provided.

This sort of motoring, calling for little exercise on the part of the driver, really shows up the comfort or otherwise of a vehicle. Far too frequently, after an hour or so, neck-ache begins, or there is an almost irresistible impulse to wriggle one's accelerator

CALM BEFORE STORMING.—The Zephyr in a rainwashed, blossom-laden setting at Genevreuille between Vesoul and Belfort before linking up with the competitors.

be for a Carré de l'Est cheese of extreme ripeness served to us with an excellent, if simple, meal. I thoroughly enjoyed it, but Moore said he wouldn't touch the stuff with a barge pole.

Next morning after a pleasant run through rain-washed countryside and between tall, blossom-laden trees, we arrived in Belfort in time to lay in provisions for the next 24 hours. The inhabitants of this town, not far from the Rhine, have obviously witnessed so much excitement and undergone such vicissitudes with the passage of armies in various wars, that the descent upon them of 200-odd rally cars appeared to cause little stir. Indeed, when the Vredestein tyre lorry drew up at the control and preparations were made for re-tyring such cars as needed it, the only spectator showing any real signs of interest was an ancient *abbé* who stood, breviary in hand, gazing through thick spectacles with a look of wonderment. There were almost as many people admiring *le Ford Anglais* as there were watching the rally preparations.

From Belfort we roared on with the rally through the Juras, the Savoy Alps and the eastern fringe of the Massif Central, through St. Claud, over the special section between here and Morez, where we managed an even better average than was required for our class, had we been competing; to the control at Champagnole and on through Aix les Bains, pulling off the road to watch groups of cars as they sped by. Whenever we ran out of competitors, short-cuts to catch them up again were found with the aid of Mr. Michelin's excellent maps. Once, however, short-cutting over an obscure by-road, we lost ourselves in the mountains, finishing up with headlamps illuminating nothing but a vast expanse of water. Quite a lot of time was thereby wasted before we got back to our proper route.

IN MEMORIAM.—This impressive piece of statuary, to commemorate the Maquis of the Ain who lost their lives during the German occupation, is being erected on the road between Bourg en Bresse and Nantua.

foot, or one's posterior has frequently to be hitched back into its proper place on the seat. We encountered none of these unpleasantnesses in the Zephyr. The bench-type front seat, with divided back squab at a fairly upright angle, possessed such height and degree of adjustment as to provide a completely relaxed position and driving for considerable distances produced no marked fatigue.

Only a couple of gallons of fuel were left in the tank after 178 miles of fast motoring; this signified no unduly heavy consumption—the average worked out at just over 25 m.p.g.—but served to emphasize the need for a tank of more than 9 gallons capacity on a car so patently designed for going far, fast. More annoying still was when, because the car was standing on a fairly steep camber, many francs-worth of fuel gushed from the almost horizontal filler pipe before the gauge showed "full." After that, unless the Ford was on a very even keel, orders for fuel were in specific quantities instead of *complet*.

As darkness fell, deteriorating roads and increasingly heavy rain reduced our cruising speed; nevertheless it was only 9.20 p.m. when we arrived at Vesoul—an average from Paris, including all stops, of 42½ m.p.h. The Hôtel du Nord, where we lodged that night, proved a comfortable little establishment of no great pretensions unless it

The thin, pre-dawn half-light was just percolating through low clouds south of Chambéry as we climbed the Col du Granier with visibility less than 10 yards, keeping our fingers crossed against anyone coming down in the opposite direction and casting timorous glances at a ghostly wooden fence outlined on our right against nothing.

And so over mountain roads to Valance, where the route turned north again, scrambling around hairpin after interminable hairpin, using engine power in second gear to augment the somewhat restricted steering lock of the

Zephyr. Then down and down and round and round, as quickly as possible so as not to impede competitors, all the time expecting the brakes to fade suddenly into uselessness under such treatment. But no such untoward thing occurred, and unlike some of the competitors we pulled in that evening at Bourg en Bresse, weary but unscathed after 26 hours of hard and more or less continuous mountain motoring.

FINE FEATHERS.—Among the birds in the garden of the hotel at Bourg are some magnificent golden pheasants. This one consented to pose for us, undeterred by the dog on the other side of the fence.

Birds Before Breakfast

The Hôtel de l'Europe at Bourg is a pleasant old coaching house where the beds are comfortable, the sanitation modern and they do you well enough to justify a cooking star in the Guide Michelin. (The local speciality, a large sort of capon bred in the district, is served in a variety of ways under the generic title of *poulet de Bresse*.) It was, nevertheless, a somewhat unnerving experience to throw open one's shutters in the morning and meet the stern gaze of a grey stork with a golden pheasant in attendance, which gave rise to a feeling that the previous evening must have been less abstemious than one remembered. However, the storks and pheasants are no hallucination and after breakfast, before leaving for Geneva, we were made free of the garden where these birds roam at will.

The only halt between Bourg and the Swiss frontier was to admire the enormous piece of statuary being erected at a corner of the road between Bourg and Nantua. Its location has obviously been chosen with great care and the female figure gazes out across the surrounding countryside where Maquis of the Ain Department, which it commemorates, lost their lives in resisting the German invaders.

With Moore on his way to Italy, the journey back to Dunkirk became a single-handed affair. The lack of an additional ratio between second and top made the climb up from Switzerland towards Morez, across mountains and, between banks of snow which, even in early May, still lined the road, a somewhat time-wasting business. Once down on more level country, however, the Zephyr again came into its own and mile after mile was reeled off at a very smart pace, the speedometer on occasion going almost up to the 90 m.p.h. mark. Duffle-coated and with the heater providing foot-warmth, this was really pleasant "top-down" motoring despite a nip in the air, for the convertible "open" is remarkably draught-free.

All through the long afternoon we sped, and through the evening, stopping on occasion for fuel and, at other times, for personal fortification. Nightfall saw us just south of St. Dizier and the clocks of Rheims were striking 10 as Mme. Morel greeted us at her very pleasant little Welcome Hotel. In $7\frac{3}{4}$ hours, including stops which cannot have amounted to less than 45 minutes, 284 miles had been left behind; total for the day, 357.

Fulfilment of editorial commitments, plus a little shopping, took up much of the following day; consequently, it was after 6.30 p.m. when the Zephyr turned its boot on the cathedral towers and set out on the last 160-mile leg to Dunkirk.

SPANNING THE CENTURIES.—The modern lines of the Zephyr in its fully-open state contrasted strongly with the Gothic magnificence of Rheims cathedral, generally regarded as the finest of its type in France.

LAY BY.—Wherever possible, halts were made in the special sections to watch the competitors. In this photograph, the scene is on the Col de Grimone which was a timed climb.

As many readers will know, the way gets worse as the road goes north and the last 50 miles or so are over extremely rough *pavé*. In the normal way one would have tackled this at reduced speed, but time was of the essence and the unfortunate Ford was forced to take it and like it: that it did not actually dislike it was gathered from the very little disturbance felt in the driving seat, although quite a lot was obviously happening underneath which subsequently showed itself by some tiredness of the shock absorbers. Thus it was that long before it need have been, this "eater-up of unconsidered kilos" was safely bestowed in the ferry boat's garage, having covered slightly over 1,700 hard miles in five days.

ROAD TEST

Ford ZEPHYR Convertible

MOTOR *Life* Test Staff Report

WITH MORE and more small "family-type" imported cars making their appearance in urban areas (See *The Race for Home*, this issue) MOTOR *Life* test staff decided to take a close look at one such import, Ford's luxury family car from Great Britain, the Zephyr. The Zephyr is the largest and most powerful of Ford's British products, the Anglia, Prefect, Consul, Zephyr and the ultra-deluxe Zephyr model, the Zodiac. In appearance, the Zephyr looks like a scaled down '49-'51 Ford with a grille which might have come directly out of the costly British Aston-Martin sports car. The designers have wisely reduced wheel size to 13" for balance in appearance and our test car, a cream-colored convertible which was loaned to us through the courtesy of Dick Scatchard of Import Motor Sales in Long Beach, California, had the addition of a continental kit spare tire assembly.

From the front, the Aston-Martin type grille is impressive and the chrome content has been nicely balanced with painted trim. The front bumper has deliberately been placed high, giving the front end the appearance of being complete with belly pan—and giving the car much-needed protection from the too-high bumper location of many domestic products. The hood ornament is referred to at the British factory as the "Dying Duck," because of its downspread wings. But the designers are quick to point out, in answer, that the "wings,"

British Ford Zephyr looks like a scaled-down 1949-1951 American Ford Car, has conventional shift, 68 horsepower six cylinder engine.

which attach solidly to the hood at their extremities, form air passageways which deflect road dust, dirt, insects and even rain, up and over the windshield. And it works. Although the cream-colored Zephyr rapidly acquired a coat of grime during the road test proceedings, the windshield stayed remarkably clean.

The windshield is high, high enough to protect drivers of normal proportions, up to 6′ 5″, before the head protrudes, a common weakness in many of the "small" cars. Front wind wings are stationary, as is the custom on convertibles made in England. The side windows roll up, and the rear quarter windows in our test convertible were erected by hand. The fit was snug when erected, and they do follow the flush body lines when lowered. The top is made of a washable plastic material, is electrically operated and can be raised to full convertible or true Coupe de Ville positions. The fit is tight around the sides, at the windows and at the juncture of top and windshield. It can be raised easily by one person and the average time for raising the top completely, including removal of the boot, ran two minutes for the test staff.

The inside of the convertible was done in a matching cream color simulated leather with cream door panels, dashboard and even ash trays, to match. The leather was trimmed with green, the same shade as the convertible plastic top. Front seats are adjustable back and forth, but perhaps too much forth and not enough back. It is doubtful that any driver would want to sit close enough to the wheel to justify the amount of travel forward and the six footers on the test crew felt that another several inches of leg room would be needed for complete safety and comfort. The instruments are grouped in one steel pressing directly ahead of the driver and visible through the spring-spoked steering wheel. The speedometer reads to 90 mph; a fairly accurate estimate of the car's top speed in conventional form. Panel is complete with gas gauge

Performance and Specifications
Ford Zephyr Convertible (British)

SPEEDOMETER ERROR

Indicated 30 mph	28.8 mph actual
Indicated 40 mph	38.2 mph actual
Indicated 50 mph	47.9 mph actual
Indicated 60 mph	57.1 mph actual
Indicated 70 mph	66.3 mph actual
Indicated 80 mph	75.4 mph actual

ACCELERATION

0-30 mph	4.9 sec.
0-45 mph	12.2 sec.
0-60 mph	21 sec.
Standing start ¼ mile	21 sec.

TOP SPEED

Fastest one-way	81 mph
Slowest one-way	75 mph

FUEL CONSUMPTION

(City, country and test run driving) ... 22.4 avg. mpg

ENGINE—In-line ohv six cylinder. Bore and stroke: 3.125 x 3 inch (Oversquare). Compression ratio 6.8:1. Displacement 2262 cubic centimeters or 138 cubic inches. Advertised horsepower 68 at 4000 rpm. Max. torque at rpm 112 at 2000.

LIST PRICE—$2345 at ports of entry.

REAR AXLE RATIO—Conventional transmission only ... 4.375

DIMENSIONS

Wheelbase	104 inches
Tread	50 in. Front; 49 in. Rear
Weight (Shipping)	2755 lbs.
Steering ratio	13.6:1
Turning radius	40 feet, 6 inches
Steering, turns lock to lock	2⅔

Convertible top is automatic, can be raised to fully closed or Coupe de Ville positions. Continental kit and extension are optional extras.

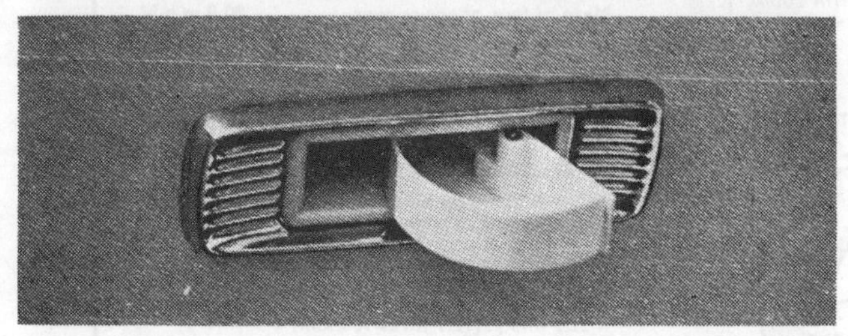

Interior features unusual extras, like swiveling plastic ashtray which hides in seat back. Upholstery is grained simulated leather.

Instrument cluster is neat, easy to read between spring-spoke steering wheel. Lever on horn button controls turn signals.

Back quarter windows are manually operated, slide in and out of well. Note how chrome window lip covers top for protection.

and temperature gauge plus warning lights for oil pressure and battery discharge. Method of starting the car is to turn the key on the left of the instrument cluster, press the starter button on the right of that same cluster and step on the overhead-suspended throttle.

A wide-mouthed dashboard shelf runs the full width of the cockpit and is handy for maps, and other paraphernalia found in the glove compartment of most cars. Unfortunately, most Americans prefer to have a compartment they can close and lock and no provision has been made for this, here. Turn signal controls are located on a button in the center of the horn ring and a cane-type parking brake pulls out of the dash, directly to the right of the steering column. Hood lock release is still located to the extreme right of the dash panel; undoubtedly the same location as in the right hand drive models of the Zephyr.

The shift is placed on the column. The clutch, as well as the brake, is of the "hanging" type pedal, as introduced on Ford cars here last year, and the clutch is hydraulic. Shift pattern is conventional American with three speeds forward and one in reverse. Shift throw is short and snap-shift technique is quite simple in this smaller Ford. In initial "snap," the Zephyr moves out with the best of them, gets to 30 mph in under 5 seconds and to 60 in 21 seconds. This compares favorably with most 1954 low-priced American cars. The best shift points, for maximum acceleration, worked out at 26 mph from first to second, 52 mph from second to third.

Even during the strenuous road test activities on the Ford Zephyr model, the gasoline mileage worked out to a handsome average of 22.4 miles per gallon. This figure included stop-and-go city driving, highway cruising and burning-rubber acceleration tests and top speed runs. We made six top speed runs, three with and three against the prevailing winds. Average of the top speed runs for this stock-engined convertible was 77 mph, with one run netting us a top speed of 81 mph. During the high speed runs, the car held true to the road, ran straight and firm against the roadbed. The only variation in direction came when the wind fishtailed and blew, momentarily, broadside. It was enough, the Zephyr yawed to the left and the test driver tugged hard on the wheel to get the car back in line. It should be mentioned, in all fairness, that the staff station wagon on the way home, "wandered" under the same circumstances.

The Zephyr ride is good; a compromise between the soft springing of American products and the very hard ride of the foreign sports cars. Unlike most short wheelbase cars, the Zephyr offers up no pitch and roll motion. The convertible, because of its additional weight, hugs the ground even more securely than the sedan.

The steering, with 2⅔ turns lock to lock, is quick and sure and the Zephyr can be handled like a sports car, in and around traffic, through gradual or deep bends and turns. There is a noticeable amount of tire squeal but the tires hold firm and the rear end of the car broke out only once, when the wheel was turned hard over at 45 mph on a sand and rock roadbed. By allowing the steering wheel complete freedom, the car returned to a straight line immediately. The turning circle of the car is a few inches over 40 feet, quite large for such a small wheelbase car. Brakes are excellent with little pressure required for "panic" stops.

Because the car looks like a scaled-down Ford, it attracts a great deal of attention. Many drivers want to know exactly what kind of a car it is, how much it costs and where they can "see" one in detail. When it takes off from the stop lights ahead of the automatic transmission models, they become even more impressed.

The Zephyr, summed up, is a breeze. It's a breeze to drive, to park, to handle, in the mountains or on the desert. Gasoline mileage is excellent for the performance and the styling will stay current for several years to come. One test staffer is purchasing the sedan model Zephyr as a second car for his family. Can we say more? •

The Motor Road Test No. 16/55

Make: Ford **Type:** Zephyr Zodiac

Makers: Ford Motor Co., Ltd., Dagenham, Essex

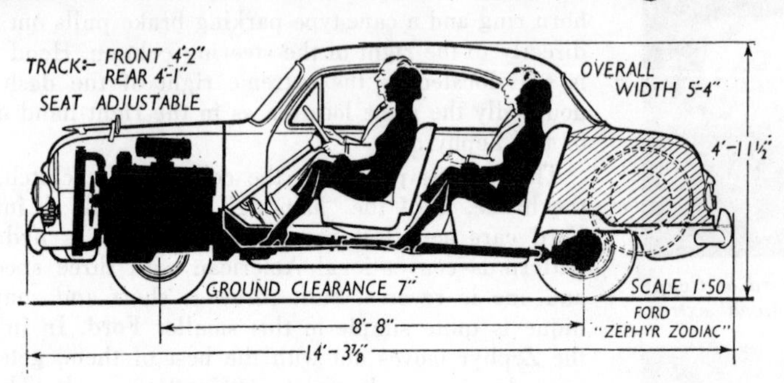

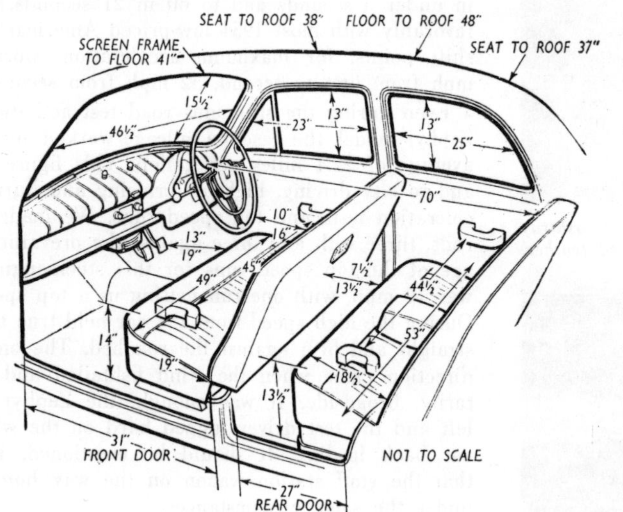

Test Data

CONDITIONS. *Weather: Warm (temperature 53°-61°F., barometer 29.4-29.5 in.), very strong wind blowing along course. Surface: Dry tar macadam. Fuel: Premium grade petrol.*

INSTRUMENTS

Speedometer at 30 m.p.h.	10% fast
Speedometer at 60 m.p.h.	2% fast
Distance recorder	2% fast

MAXIMUM SPEEDS

Flying Quarter Mile

Mean of four opposite runs	80.0 m.p.h.
Best time equals	85.7 m.p.h.

Speed in gears

Max. speed in 2nd gear	53 m.p.h.
Max. speed in 1st gear	30 m.p.h.

FUEL CONSUMPTION

32.5 m.p.g. at constant 30 m.p.h.
30.5 m.p.g. at constant 40 m.p.h.
26.5 m.p.g. at constant 50 m.p.h.
23.0 m.p.g. at constant 60 m.p.h.
20.0 m.p.g. at constant 70 m.p.h.
Overall consumption for 895 miles, 40.25 gallons, =22.2 m.p.g. (12.7 litres/100 km.)
Fuel tank capacity 9 gallons.

ACCELERATION TIMES Through Gears

0-30 m.p.h.	5.6 sec.
0-40 m.p.h.	9.4 sec.
0-50 m.p.h.	14.1 sec.
0-60 m.p.h.	20.2 sec.
0-70 m.p.h.	30.1 sec.
Standing Quarter Mile	21.8 sec.

ACCELERATION TIMES on Two Upper Ratios

	Top	2nd
10-30 m.p.h...	8.7 sec.	5.4 sec.
20-40 m.p.h...	8.8 sec.	6.1 sec.
30-50 m.p.h...	9.1 sec.	8.7 sec.
40-60 m.p.h...	11.3 sec.	—
50-70 m.p.h...	16.1 sec.	—

WEIGHT

Unladen kerb weight	23¼ cwt.
Front/rear weight distribution ..	60/40
Weight laden as tested	26¾ cwt.

HILL CLIMBING (at steady speeds)

Max. top gear speed on 1 in 20	70 m.p.h.
Max. top gear speed on 1 in 15	62 m.p.h.
Max. top gear speed on 1 in 10	51 m.p.h.
Max. gradient on top gear	1 in 7.6 (Tapley 290 lb./ton)
Max. gradient on 2nd gear	1 in 4.9 (Tapley 455 lb./ton)

BRAKES at 30 m.p.h.

0.95g retardation ..	(= 31½ ft. stopping distance) with 95 lb. pedal pressure
0.65g retardation ..	(= 45½ ft. stopping distance) with 75 lb. pedal pressure
0.45g retardation ..	(= 67 ft. stopping distance) with 50 lb. pedal pressure
0.14g retardation ..	(= 215 ft. stopping distance) with 25 lb. pedal pressure

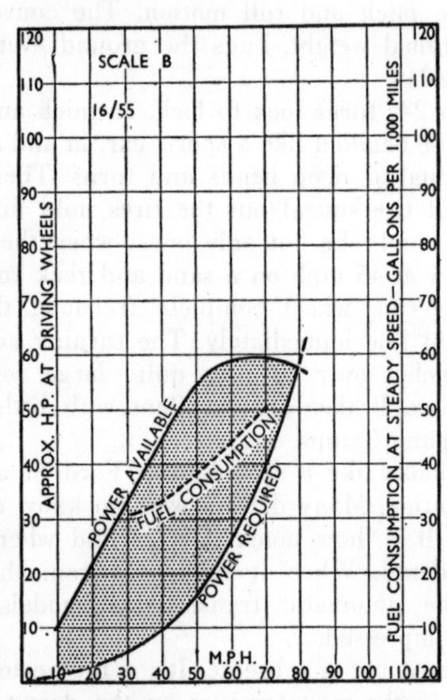

Drag at 10 m.p.h. .. .. .. 36 lb.
Drag at 60 m.p.h. .. .. .. 160 lb.
Specific Fuel Consumption when cruising at 80% of maximum speed (i.e. 64 m.p.h.) on level road, based on power delivered to rear wheels .. .. 0.76 pints/b.h.p./hr.

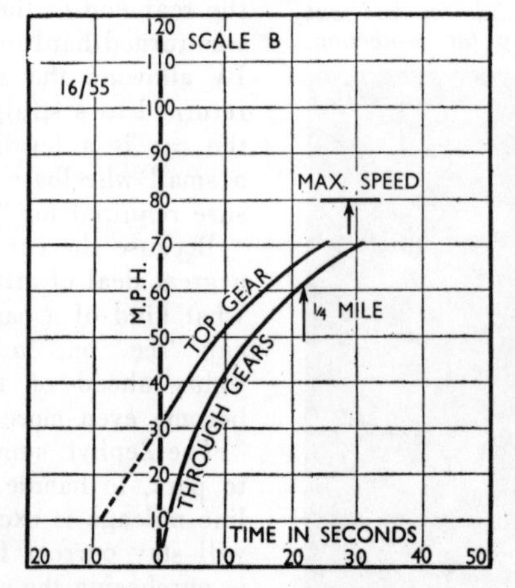

Maintenance

Sump: 6½ pints (plus 1½ for filter), S.A.E. 20. **Gearbox:** 2½ pints, S.A.E. 80. **Rear axle:** 2½ pints, S.A.E. 90 hypoid. **Steering gear:** ½ pint, S.A.E. 80 E.P. **Radiator:** 22 pints (2 drain taps). **Chassis lubrication:** By grease gun every 1,000 miles to 14 points. **Ignition timing:** 15° B.T.D.C. **Spark plug gap:** 0.032 in. **Contact breaker gap:** 0.014/0.016 in. **Valve timing:** Inlet opens 17° B.T.D.C. and closes 51° A.B.D.C. Exhaust opens 49° B.B.D.C. and closes 19° A.T.D.C. **Tappet clearances:** (Cold) Inlet and exhaust 0.014 in. **Front wheel toe-in:** 0-⅛ in. **Camber angle:** + 2°/—½°. **Castor angle:** +½°/—½°. **Tyre pressures:** Front 24 lb. Rear 24 lb. **Brake fluid:** Enfo ME-3833-C. **Battery:** 12 volt, 51 amp./hr. **Lamp bulbs:** Headlamps, 42/36 watt; side/direction indicators, 18/6 watt; stop/tail lamps, 18/6 watt; rear direction indicators, 18 watt; rear number plate lamps, 6 watt; pass light and long-range driving lamps, 48 watt; reversing lamp, 18 watt; roof lamp, 6 watt; instrument panel and warning lamps, 3 watt; reversing lamp switch warning lamp, 1.5 watt; clock and cigar lighter, 2.2 watt.

Ref. B/23/55.

The Ford Zephyr Zodiac

Special Equipment Gives Added Appeal to an Already-popular 2¼-litre Saloon

In Brief

Price: £600 plus purchase tax £251 2s. 6d. equals £851 2s. 6d.

Capacity	2,262 c.c.
Unladen kerb weight ...	23¼ cwt.
Fuel consumption	22.2 m.p.g.
Maximum speed	80 m.p.h.
Maximum speed on 1 in 20 gradient	70 m.p.h.
Maximum top gear gradient	1 in 7.6
Acceleration:	
10-30 m.p.h. in top ...	8.7 sec.
0-50 m.p.h. through gears	14.1 sec.

Gearing: 16 m.p.h. in top at 1,000 r.p.m.; 80 m.p.h. at 2,500 ft. per min. piston speed.

ALTHOUGH painting the lily is rightly held in mild contempt, a somewhat similar process can be applied to modern quantity-produced cars with results that are both sensible and pleasing. This state of affairs arises from the fact that quantity production offers a number of fundamental advantages which, in the final result, add up to a highly satisfactory product at a very competitive price—with, however, the qualification that the buyer must accept the standards of finish and equipment appropriate to the widest possible sale of the model concerned.

Special Edition

A mild attempt to bridge the gap between such cars and considerably more expensive types of higher quality and smaller production has existed for many years, in the shape of better-equipped de luxe editions of standard models, and the ever-extending plan of making various extras available to the buyer who wants them. It remained to Ford Motor Co. Ltd., however, to foresee the full implications of modern demands in this respect and to consolidate the requirements of such buyers into a separate model with a de luxe look as well as many de luxe features. The Zephyr Zodiac, introduced at the Earls Court Motor Show of autumn 1953, was the outcome, and it is superfluous to remark that it met with instant success.

Principal differences between the Zephyr Zodiac and the standard Zephyr Six include an attractive range of two-tone paintwork matched by two-tone leather upholstery, the fitting of whitewall tyres and the inclusion of such items of equipment as a heater and demister (extra on the normal Zephyr), screen washer, extra rear-view mirrors on the front wings, larger battery, cigar lighter, clock, reversing light and that most excellent combination for the fast night driver, the large-size Lucas fog light and its mate, the long-range driving lamp (familiarly known as the "flame-thrower").

In addition, the six-cylinder Zodiac engine has a compression ratio of 7.5 to 1 (standard Zephyr, 6.8 to 1), which takes advantage of premium-grade fuel to put up the power output by 3 b.h.p. to 71 without increase in peak speed (4,200 r.p.m.).

As one would expect, this increase in power output of approximately 4% results in a useful rather than startling all-round improvement in performance. Before going on to make comparisons with a convertible example of the normal Zephyr Six tested last November, however, it should be stressed that weather conditions for the Zodiac test were distinctly unfavourable in that a strong wind was blowing along the course. Given better conditions, it is probable that the difference between the Zodiac and the standard models would be even more marked.

Despite this fact, the Zodiac showed an improvement in acceleration time from 10 m.p.h. to 70 m.p.h. in top gear of 10.6%, the improvement, as one would expect, being most marked at higher speeds. Thus a gain of 4½% in the 10-30 m.p.h. time increased to just on 14% in the 50-70 m.p.h. figure, which represents a most valuable benefit for the fast driver.

In fuel consumption, the constant-speed recordings show gains of around 3% at most speeds and considerably higher than that (just over 8%) at 70 m.p.h. Here again the higher compression engine showed some benefit for all drivers and an appreciable benefit for those who habitually travel fast. So far as the overall consumption figure is concerned, the figure obtained is worse than many drivers would obtain, as the car was driven hard throughout most of the period it was in our hands.

On premium fuel (which the manufacturers specifically recommend for this model) the engine is both smooth and free from pinking or running on. Indeed, the Zodiac is a particularly attractive car in and around town, where its quiet running but vivid response to the throttle at quite low speeds make it an unusually satisfying vehicle for business usage or shopping

TWO TONE paintwork is available on the Zephyr Zodiac, more elaborately equipped version of the Zephyr Six, with a high-compression engine to take fuller advantage of premium-grade fuel. Matched long-range and wide-beam additional lamps, external driving mirrors and white-wall tyres are other details evident in this picture.

expeditions. Road noise, which is an annoying feature of many otherwise-quiet cars, is almost entirely absent.

On the open road, the engine remains smooth throughout its range and cruises easily in the "sixties" and "seventies," although some trace of manifold roar becomes noticeable as 60 m.p.h. is approached. Above 70 m.p.h. there is, in addition, a rumble which, one imagines, has its origin in the transmission rather than in the engine but which does not prevent full use being made of the extremely lively performance.

The fact that it takes a hill of appreciably steeper gradient than 1 in 8 to enforce a change into second gear speaks for itself in the matter of hill-climbing abilities. It also emphasizes the considerable overlap which exists in the performance of this car in its various gears, despite the use of a three-speed gearbox with fairly wide spacing of the ratios. In the middle ratio, 50 m.p.h. can be exceeded but, for all normal purposes, 40 m.p.h. may be regarded as a comfortable changing-up speed for brisk results, the acceleration between 40 m.p.h. and 50 m.p.h. in top being very nearly as good as in second.

Starting properties are good and the choke can soon be dispensed with, but drivers who like to run their cars out of the garage and then leave them idling at a fast tick-over speed whilst they close the doors will find the Zodiac slightly irritating in that there is no provision for leaving the control set for a fast idle and the engine tends to stall if the control is allowed to return fully.

Engine accessibility is of a high order, all the points requiring routine maintenance being very easy to reach. No water was required during the thousand-odd miles of the test, and little over one pint of oil was needed to restore the level of the sump at the end of this distance.

FOUR-LIGHT bodywork gives a good view out of all seats, the doors being large, and the curved-glass rear window wide and deep.

Arrangements for fuel replenishing are not ideal as the filler intake is horizontal and the neck is then taken round a sharp bend, with the result that a petrol pump nozzle has to be inserted at exactly the right angle if the tank is to be filled quickly without risk of spilling. Those who travel long distances would also like to see a rather bigger tank as the present 9-gallon capacity calls for refuelling stops at intervals of less than 200 miles.

The very versatile performance of this model and its capacity for reaching high cruising speeds quickly and maintaining them well, makes good suspension and handling qualities an essential. In these respects the Zodiac does not disappoint. The springing is moderately soft but at all times well controlled. There is no pitching and ordinarily very little roll, the car cornering extremely well within and beyond normal touring limits. If cornered really drastically, some roll is induced and the small proportion of weight on the rear wheels then becomes noticeable in that the tail will hop sideways slightly on a dry road or begin to slide on a slippery surface, but a quick twitch of the steering wheel rapidly restores the status quo. As one would expect, wheelspin can readily be induced by harsh acceleration in first or second gears on a slippery surface. The car proved more susceptible to the effects of side winds than some and the steering was also affected to a minor degree by the longitudinal ruts or ridges found on some concrete roads which have a central join between sections of concrete.

These minor faults apart, steering is of a high order, combining fairly light operation with moderately high gearing; bad surfaces produced no annoying reaction through the wheel and the car can be placed accurately with confidence. The steering lock could, however, be improved with advantage.

In both the placing and operation of controls, the Zodiac showed evidence of an obvious appreciation of drivers' needs. The steering wheel, with its half horn ring and T-arrangement of the spokes is of a size and disposition to give easy control, and the steering column gear-change is normally light in action with a quite moderate travel on the lever. The synchromesh on the two upper ratios is effective, although this and the spacing of the ratios

INTERIOR layout is shown in these two photographs, details being the easily adjusted bench-type front seat, instrument nacelle facing the driver, full-width parcel shelf, door-mounted armrests, and hanging-type pedals.

BEHIND the spare wheel a tool box is provided. Reached through a spring-counterbalanced lid, the luggage locker is roomy and of convenient shape.

discourages attempts at ultra-fast gear changes, and although first gear is unsynchronized, it has a dog-clutch engagement which is quite easy at low traffic speeds—second gear suffices for virtually any hill likely to be met by tourists.

The pedals are well spaced and all are of the hanging type, the clutch, as well as the brake, having Girling hydraulic operation. An unusual but quite useful addition is a rest for the right foot designed to make it easier to maintain a steady pressure on the accelerator on long fast stretches.

The brakes provide good stopping powers with moderate pedal pressures, but on one or two occasions the foot brake produced some front-end judder at speeds below 10 m.p.h.

Of the pistol-grip type, the hand brake has adequate power and is nicely placed in relation to the gear lever for a quick get-away; indeed, the Zodiac is particularly well arranged in this respect as the ignition switch is placed to the left of the instrument binnacle and the starter-motor switch to the right, so that switching on, starting the engine, engaging a gear and releasing the handbrake all follow a natural and quick sequence of movements.

The remaining minor controls deserve praise, for being logically placed where they are easily reached and not liable to confusion instead of being arbitrarily grouped together. Less pleasing are the instruments, in the design of which styling has been allowed to overrule clarity. The arc-type speedometer, for example, is nicely in the range of vision, but the "Five Star" theme has been introduced to identify the "30" to "50" markings, making an already-indefinite scale much more so. In addition, the fuel gauge for the 9-gallon tank has five irregularly-spaced dots and, as five into nine won't go, the readings are a matter of speculation rather than precision.

In the matter of all-round vision, one would find it hard to fault the Zodiac and the good window area shows to even better advantage by reason of a seating position which brings the eye level to a point ideally suited for the best view. The usual tinted interior rear-view mirror has a very satisfactory range and the special wing mirrors completely eliminate the blind spots inevitable with a single interior fixture. In addition they serve a very useful purpose as width indicators although, strangely enough, they produce a curious illusion of extra width until one has become accustomed to them. The special spring mounting by which an accidental knock does not derange the adjustment is both ingenious and effective.

The screen wipers clear a large area and, although of the vacuum type and to some extent susceptible to effects of throttle opening, they continue to work at an adequate speed even when the accelerator is fully depressed. A much appreciated Zodiac extra is a screen washer.

The bench-type front seat is deeply sprung, has a particularly easy adjustment and is schemed to give an alert position for the driver. There is no central armrest but fixed armrests are provided on both front doors. At the rear, the seats are deep and comfortable with fixed armrests on the doors and a folding central armrest, whilst a cut-away at the base of the front seat ensures plenty of foot room.

For ventilation there are the usual hinged panels on the front door and, in addition, there is a fresh-air vent by which a ducted supply can be admitted above the foot ramp of the passenger's side. The heating and demisting system also admits fresh air, either hot or cold, but although the amount of heat provided is sufficient for moderately cold weather, the hot air is admitted above the pedals, so that a passenger who suffers from cold feet is not catered for very adequately.

Provision for luggage and odds and ends is good, with useful parcel shelves below the scuttle and a boot which provides plenty of suitcase space as well as having a convenient tool locker adjacent to the spare wheel, plus handy spaces for incidental items in the wing recesses.

In all, this Zodiac represents a most attractive answer to the needs of motorists whose basic requirements are catered for by the normal Zephyr Six but who want (and are prepared to pay a little extra for) that added touch of luxury which makes such a difference to the pleasure of both usage and possession.

Mechanical Specification

Engine

Cylinders	6
Bore	79.37 mm.
Stroke	76.20 mm.
Cubic capacity	2,262 c.c.
Piston area	46.1 sq. in.
Valves	Overhead (push rod)
Compression ratio	7.5/1
Max. power	71 b.h.p. at 4,200 r.p.m
Piston speed at max. b.h.p.	2,100 ft. per min.
Carburetter	Zenith downdraught
Ignition	12-volt coil
Sparking plugs	Champion N8B (14 mm.)
Fuel pump	AC mechanical
Oil filter	Full-flow

Transmission

Clutch	Single dry plate (hydraulic control)
Top gear (s/m)	4.444
2nd gear (s/m)	7.297
1st gear	12.62
Propeller shaft	Hardy Spicer open
Final drive	Hypoid bevel
Top gear m.p.h. at 1,000 r.p.m.	16.0
Top gear m.p.h. at 1,000 ft.min piston speed	32.0

Chassis

Brakes	Girling hydraulic (2 l.s. front)
Brake drum diameter	9 in.
Friction lining area	121 sq. in.
Suspension:	
Front	Coil i.f.s.
Rear	Semi-elliptic
Shock absorbers:	
Rear	Piston-type hydraulic
Front	Telescopic incorporated in i.f.s.
Tyres	6.40—13

Steering

Steering gear	Burman worm and peg
Turning circle (between kerbs):	
Left	44½ feet
Right	39 feet
Turns of steering wheel, lock to lock	2¾

Performance factors (at laden weight as tested):

Piston area, sq. in. per ton	34.5
Brake lining area, sq. in. per ton	90.5
Specific displacement, litres per ton mile	3,190

Described in *The Motor*, October 21, 1953

Coachwork and Equipment

Bumper height with car unladen:	
Front (max.) 22½ in., (min.) 13 in.	
Rear (max.) 24½ in., (min.) 15 in.	
Starting handle	None
Battery mounting	On left of engine
Jack	Tubular bipod type
Jacking points	2 on body sides

Standard tool kit: Wheelbrace, 2 box spanners, jack, grease gun, pliers, 2 screwdrivers, tyre lever, 2 double ended spanners, plug spanner, adjustable spanner, tommy bar and brake adjuster. (Export models, tyre pump).

Exterior lights: 2 headlamps, 2 sidelamps/direction indicators, passlamp, long-range driving lamp, 2 stop/tail/direction indicator lamps, 2 rear number plate lamps, reversing lamp.

Direction indicators	Flashing type, self-cancelling

Windscreen wipers: Dual self-parking

Sun vizors	2

Instruments: Speedometer (with non-trip decimal mileage recorder), fuel gauge, ammeter (or thermometer if requested for export) electric clock.

Warning lights: Dynamo charge, oil pressure, main beam, direction indicator.

Locks:	
With ignition key	Driver's door and boot
With other keys	None
Glove lockers	None
Map pockets	None
Parcel shelves	Under facia and behind rear squab
Ashtrays	Two (one in facia and one in back of front seat)
Cigar lighters	One (on facia)
Interior lights	One (roof)
Interior heater	Smith's fresh-air type (with demister)
Car radio	Optional

Extras available: Ekco radio, heater (on export models), "Enfo" accessories.

Upholstery material: Leather (two-tone) on wearing surfaces, p.v.c. elsewhere.

Floor covering	Pile carpets

In addition to items already mentioned, standard equipment includes windscreen washer, wing mirrors, locking petrol cap and chromium-plated wheel embellishers.

Exterior colours standardized: Single and dual colours, including Dorchester grey, Canterbury green, Bristol fawn, Winchester blue, Westminster blue, Lichfield green, black.

Alternative body styles	None

Several changes have been made in the appearance of the Consul since its introduction. Deep, wrap-round bumpers are fitted, and a chromium strip extends across each side of the car. A motif is fitted on the bonnet

The Autocar ROAD TESTS

FORD CONSUL SALOON

MANY detail changes have been made in the appearance and mechanical specification of the Ford Consul since the British component of the Ford organization introduced it at the London Show in 1950, and since a Road Test appeared in *The Autocar* on April 13, 1951. Particular interest centred on the Consul, and its larger-engined brother, the Zephyr, upon their introduction, because both models were a complete breakaway from Ford practice of the past, and it was clear that the cars would be in production for a long period to justify tooling up for such a radical change in design.

It follows that while the basic conception remains unchanged, many details have been improved in the process of development, based on experience of the car on the road, and to meet customer requirements.

Differences in the exterior compared with the car tested in 1951 include a horizontal chromium-plated strip running the length of each side of the car, chromium surrounds to the windscreen and rear window, and a bonnet motif. The improved bumpers wrap round farther, and are of deeper section. The front bumper now has a flattened portion against which the number plate is mounted, and the optionally extra over-riders are of a different section. The incorporation of rear reflectors has changed the shape of the tail light assemblies, which additionally house the flashing type indicators that have replaced the semaphore signals. The body sealing has also received attention.

The main change in the interior—which took place some time ago—is a complete restyling of the facia. The instruments are grouped under a cowl directly in front of the driver, where they are easily seen through the upper half of the T-spoked steering wheel. A wide parcels shelf has been fitted below the facia, running the full width of the car. Movement of the front, bench-type seat has been increased to six inches; it slides easily and, sensibly, rises as it moves forward. Many additions have been made to the chromium plating, including surrounds for the rear ash-tray and the interior light, and strips on the door trim. Moulded rubber flooring replaces a plastic material and, coloured to match the upholstery, it improves the interior appearance.

The engine mounting has been improved, and there are several modifications to the mechanical equipment itself, varying from chromium-plated piston top rings, an improved oil pump, and a double acting fuel and vacuum pump, to cast-iron valve tappets and improved valve stem oil sealing.

The axle ratio has been changed from 4.625 to 4.556 to 1, thus gearing the car up a little and slightly reducing the r.p.m. at any given road speed. Revised clutch disc linings have been adopted, although clutch take-up on the car tested was not specially smooth. Rubber has been introduced between the rear springs and the axle, and the silencer mounting has been altered to eliminate vibration.

There are also several changes in the electrical specification. The rather unimpressive high frequency horn that

Restyled tail lamps now incorporate reflectors and the winking indicators which have replaced the earlier semaphore type. There is a chromium surround to the windscreen and rear window. The bumper over-riders are optional extras

used to be mounted under the bonnet has been discarded in favour of a single Windtone mounted behind the radiator grille. The battery now fitted as standard, made by the Ford company itself, has a hard rubber case and Porvic separators instead of a compound case and wood separators, and the battery has a shield against engine heat. The ignition equipment includes an oil-filled coil.

The characteristics of the Consul are widely known, both to owners by experience and to prospective or would-be buyers by word of mouth. A criticism levelled at the car is directed at the weight distribution, which leads to some lack of traction on slippery surfaces, owing to the rear wheels being more lightly loaded than those at the front. This weakness exists but, like so many reports of such a nature, it has tended to become exaggerated among those who have no personal knowledge of the car. Adhesion is excellent on dry roads, good in slippery conditions; and wheelspin is likely to be encountered only when a jerky gear change is made from top to middle gear (of the three-speed box) on slippery surfaces. It is certainly not necessary to weight the luggage locker, as is sometimes suggested by those who lack actual road experience of the car.

The flat facia has been discarded in favour of a hooded mount for the instruments directly in front of the driver, clearly seen by the driver through the upper half of the T-spoked steering wheel. An adjustable foot rest is mounted to the right of the throttle pedal. The minor controls, however, are widely scattered

The Consul has considerable appeal, reasons for its popularity being readily apparent. Although of compact dimensions, the car really does seat up to six people of average size in reasonable comfort; many drivers might prefer separate front seats for reasons of support, but they have the consolation that at least the front bench-type seat, coupled with the steering column gear change, enables an extra passenger to be carried in the front when necessary.

Performance is impressive, the ride is comfortable, and the handling is such that the enthusiastic owner may, when he feels so disposed, throw the car into corners knowing that no unexpected vice will catch him unawares. In its current form the fuel consumption is satisfactory, the car tested yielding nearly 27 m.p.g., even when driven hard on routes that included a proportion of city traffic conditions, and 31 m.p.g. when driven quietly.

The badge on the radiator grille has been redesigned, and there is a motif on the bonnet. The number plate is now mounted on the redesigned bumper. The side lights incorporate winking indicators

The exterior is pleasingly smooth and the lines are among the best of current quantity produced cars. There is no unnecessarily fussy ornamentation: the car looks right. The all-important addition to these virtues is, of course, that the price is moderate in the light of the nature and capabilities of the car.

The Consul has an "over-square" engine: that is to say the piston stroke is less than the diameter of the cylinder bore. Thus, even at high r.p.m., the piston speeds are relatively low, and the engine will rev freely right up to its maximum without the kind of protest that suggests that it may "burst" at any moment. The car will cruise at around 55 m.p.h. without sign of strain, and speeds in the sixties can be held without the suggestion that the engine is being unreasonably over-worked. The maximum speed of the car under consideration was slightly less than that of the example tested in 1951, but this was very probably the result of a rather stiff cross wind. Nevertheless, the acceleration figures were better, and the top speed was well over a true 70 m.p.h., which is creditable for a quantity produced 1½-litre car with such a spacious interior.

Not everyone likes to drive fast, but it is important to

Access to front and rear is excellent, and all doors are hinged at their forward edges. Leathercloth is used for the upholstery as before, but the floor is now covered in moulded rubber. Chromium-plated embellishments have been fitted to the doors and round the ashtray in the rear seat

FORD CONSUL ROAD TEST . . .

Accessibility to engine components generally is good, although the dip stick is buried low down where it is difficult to replace in anything other than good lighting. Opposite the battery is a part of the heater system, beyond which is the fluid reservoir for the master cylinders of the brakes and clutch linkages

record the behaviour of a car at or near its maximum because it may well be the measure of the safety margin that exists at any speed, and, of course, there is little to choose between most cars at, say, a modest 40.

Right up to its maximum the Consul remains under complete control. It emerges from hard cornering holding a good line and without an unpleasant amount of roll. Always there are the powerful brakes to provide that important reserve of safety; they can be applied even on a bend without any marked change in steering characteristics, unless, of course, the speed is such that power from the driving wheels is necessary to maintain direction. The brakes are light to operate, and the hand brake is powerful, although the pull-out lever for the left hand, located under the parcel shelf, might be more accessible.

There has been some criticism of the steering of the Consul. It seems almost unfair to criticise the steering mechanism, for the action is so light, and at the speeds usually used by most drivers it is precise. But at the higher speeds, if sharp changes of direction are required in quick succession, the steering does leave something to be desired, and tends to show up the potential advantage of the rack and pinion system as opposed to the worm and peg type used on the Consul. Indeed, given really good steering, this model could be outstanding in its combination of comfort and ease of control.

The suspension is soft and comfortable without being spongy, the stability aided by the unorthodox form of independent front suspension that uses long hollow king pins and coil springs whose top ends are mounted high under the wings. It is almost indifferent to rough surfaces, which can be taken at various speeds without the occupants being jarred. There is no tendency to pitch, and vertical motion induced by bumps is well controlled.

A three-speed gear box inevitably leaves something to be desired by enthusiastic drivers, but on the Consul the shortcomings are compensated to a great extent by the unusually efficient steering column change mechanism and the good choice of ratios. Middle gear is sufficiently low to make the use of first largely unnecessary once the car is on the move, the lowest gear being required only when the driver is baulked on a steep hill, or at junctions that require a reduction of speed almost to a stop, yet it permits acceleration up to 40 m.p.h. very comfortably and up to nearly 50 m.p.h. if necessary.

The change mechanism is the more simple because a total of four instead of the more usual five gears has to be engaged. Changes between top and second can be made by using a couple of fingers of the left hand without releasing the steering wheel, which assists smooth, easy driving in town traffic. First gear must be used on the move from time to time on occasions on which second gear would suffice on a four-speed box, and it is at these times that the lack of synchromesh on this gear is noticed. To use the gear box well it is necessary to double-declutch when changing down into first, a manœuvre unknown to many "modern" drivers, and one with which women usually have difficulty.

The instruments, grouped in the cowl already mentioned, comprise a speedometer with a total—but not trip—mileage recorder, light indicators for the main head lamp beam and the direction winkers, an ammeter and a fuel gauge. There is also a light indicator instead of an oil pressure gauge. Instrument lighting at night is controlled by a separate switch that is overridden by the side lamp control, so that the panel lights cannot be left on inadvertently. There is no rheostat, but the degree of illumination is adequate without being distracting.

Double-dip head lamps are fitted, and the illumination they provide is adequate for normal night driving, although it falls short of the intensity required when a driver wishes to use fully the available performance on a night run.

The driving mirror is well sited, and provides a better than average rear view. There are two sun vizors which may be swung to the sides when necessary. In the hot weather in which for the most part the car was tested, the ventilator, with its outlets at toe-board level, was appreciated. Additionally, good ventilation was provided by the swivelling windows in the front doors, which could be parti-

A creditable amount of luggage can be stored in the locker, in spite of the space taken up by the spare wheel. Small cans, polishes, and so on, may be stored in the locker's double walls. The number plate is illuminated by a light unit mounted in the edge of the locker lid, and reflectors are mounted above the stop, tail and winking lights

ally opened for use as extractors or swung round to act as scoops.

The car tested was fitted with the optionally extra heater-demister and the radio unit. The heater takes in fresh air from a duct behind the radiator grille where the heater fan is also mounted, and there is full control of the temperature and flow of air to the interior or to the screen. Past experience of this heating system on the Consul enables it to be recorded here as effective.

When factory fitted, the radio is a neat installation on the parcels shelf. The controls can be reached by the driver or front passenger, and the speaker is also central, mounted above the windscreen. There are three manually tuned drums which, if desired, can be left tuned to the required stations, and then selected by switch. Reception and tone are excellent.

There is much about the Consul that is impressive, notably detail points of design that may not make themselves apparent under casual inspection. The windscreen wipers, for example, are really effective on the curved screen; suction operation reduces the demands on the battery, and there is no tendency for them to "dry up" in the way that the suction type did in the past. They are driven by a vacuum pump that works in conjunction with the fuel pump. Although the rear seat is well within the wheelbase to provide a comfortable ride and space for good luggage accommodation, the leg room at front and rear fully meets the requirements of the majority of human sizes.

For carrying capacity, economy, performance and modest initial price and running cost, the current Consul merits praise. Yet additionally it incorporates that air of "rightness" that raises it from the level of merely efficient and safe transport to one which earns the affection and respect of its owner.

FORD CONSUL SALOON

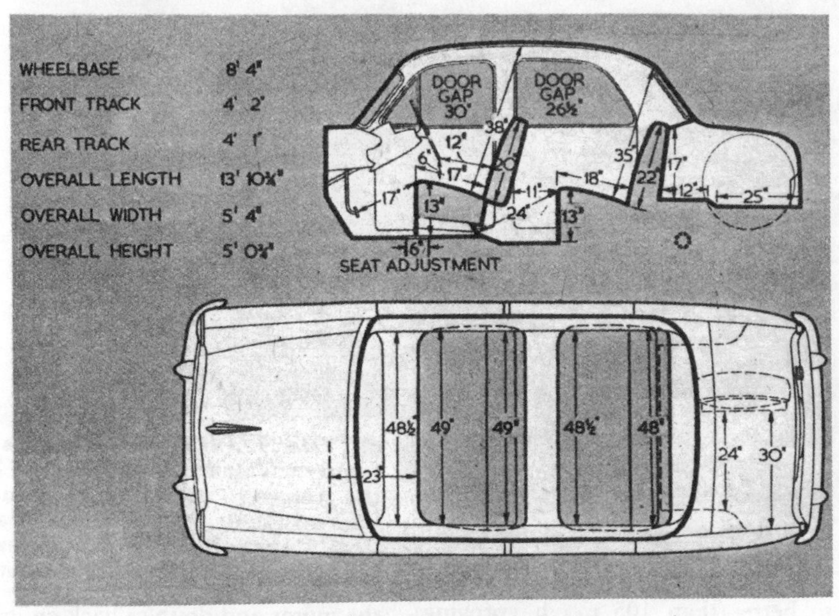

Measurements in these ⅛in to 1ft scale body diagrams are taken with the driving seat in the central position of fore and aft adjustment and with the seat cushions uncompressed

PERFORMANCE

ACCELERATION: from constant speeds
Speed Range, Gear Ratios and Time in sec.

M.P.H.	4.56 to 1	7.48 to 1	12.939 to 1
10—30	12.5	6.8	—
20—40	12.1	7.3	—
30—50	14.5	—	—
40—60	17.8	—	—

From rest through gears to:

M.P.H.	sec.
30	6.9
50	17.2
60	25.9

Standing quarter mile, 23.4 sec.

SPEEDS ON GEARS:

Gear		M.P.H. (normal and max.)	K.P.H. (normal and max.)
Top	(mean)	73	117.5
	(best)	74	119.1
2nd		40—49	64—79
1st		20—29	32—47

TRACTIVE RESISTANCE: 43 lb per ton at 10 M.P.H.

TRACTIVE EFFORT:

	Pull (lb per ton)	Equivalent Gradient
Top	240	1 in 9⅓
Second	355	1 in 6¼

BRAKES:

Efficiency	Pedal Pressure (lb)
45 per cent	25
67 per cent	50
78 per cent	75

FUEL CONSUMPTION:
29 m.p.g. overall for 420 miles (46.7 litres per 100 km).
Approximate normal range 26-31 m.p.g. (10.86-9.21 litres per 100 km).

WEATHER: Dry, sunny, cross-wind.
Air temperature 66 deg F.
Acceleration figures are the means of several runs in opposite directions.
Tractive effort and resistance obtained by Tapley meter.
Model described in *The Autocar* of October 20, 1950.

SPEEDOMETER CORRECTION: M.P.H.

Car speedometer:	10	20	30	40	50	60	70	80
True speed:	8	18	28	38	47	56	66	74

DATA

PRICE (basic), with saloon body, **£470.**
British purchase tax, **£196 19s 2d.**
Total (in Great Britain), **£666 19s 2d.**
Extras: Radio **£30 2s 1d.** Heater, **£17.** Over-riders **£4 12s 1d.**

ENGINE: Capacity: 1,508 c.c. (92 cu in).
Number of cylinders: 4.
Bore and stroke: 79.37 × 76.2 mm (3.125 × 3in).
Valve gear: o.h.v., pushrods.
Compression ratio: 6.8 to 1.
B.H.P.: 47 at 4,400 r.p.m. (B.H.P. per ton laden 39.87).
Torque: 72 lb ft at 2,000 r.p.m.
M.P.H. per 1,000 r.p.m. on top gear, 15.3.

WEIGHT: (with 5 gals fuel), 20⅝ cwt (2,310 lb).
Weight distribution (per cent): F, 60; R, 40.
Laden as tested: 23½ cwt (2,637 lb).
Lb per c.c. (laden): 1.74.

BRAKES: Type: Girling.
Method of operation: Hydraulic.
Drum dimensions: F, 9in diameter; 1.75in wide. R, 9in diameter; 1.75in wide.
Lining area: F, 60.5 sq in. R, 60.5 sq in (51.5 sq in per ton laden).

TYRES: 5.90 — 13in.
Pressures (lb per sq in); F, 28; R, 28 (normal).

TANK CAPACITY: 9 Imperial gallons.
Oil sump, 6 pints + 1½ pints in oil filter.
Cooling system, 16⅝ pints (plus 1½ pints if heater is fitted).

TURNING CIRCLE: 40ft 6in (L and R).
Steering wheel turns (lock to lock): 2⅝.

DIMENSIONS: Wheelbase: 8ft 4in.
Track: F, 4ft 2in; R, 4ft 1in.
Length (overall): 13ft 10⅝in.
Height: 5ft ¼in.
Width: 5ft 4in.
Ground clearance: 6.6in.
Frontal area: 20 sq ft (approximately).

ELECTRICAL SYSTEM: 12-volt; 45 ampère-hour battery.
Head lights: Double dip; 42-36 watt bulbs.

SUSPENSION: Front, independent, coil springs, anti-roll bar. Rear, longitudinal half-elliptic leaf springs.

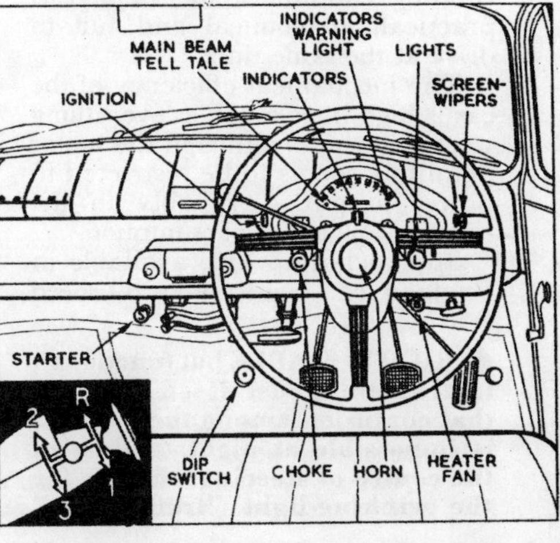

ROAD TEST

ROADWORTHY ZEPHYR

● **THE ZEPHYR for 1955 is a family car and can be handled by the family. Apart from minor improvements, appearance is practically unchanged from last year. Lines of the Zephyr are smooth.**

WITH a maximum speed of just over 80 m.p.h., nobody could say that the Ford Zephyr is a really fast 23-horsepower car. Yet this vehicle has the ability to put many effortless miles into an hour, thanks to good road holding and astounding top gear hill climbing.

Seating six people and giving better than 28 m.p.g. under normal running conditions, the Zephyr is practical, economical and fun to drive at the same time.

The mechanical efficiency of the car is high in as much as everything works very well, although the motor definitely seems to be restricted in power output — probably in the cause of low petrol consumption.

Special hot-up kits, available in England, can boost the Ford's speed to better than 105 m.p.h., proving the point that the motor has been slowed down for longer life and economy.

On the standard motor, the inlet manifold is full of right-angle bends, and the exhaust starts at the back of the motor and doubles back on itself after passing the first port.

Flat out in the two lower gears giving speeds of 30 and 50 m.p.h., produced violent valve crash for the last three m.p.h.

For normal brisk motoring, the

● **RIGHT: SIMPLE but functional, instrument panel is so laid out that confusion among the switches is impossible at night. Switch in the centre of steering wheel is for the winking-light "trafficators."**

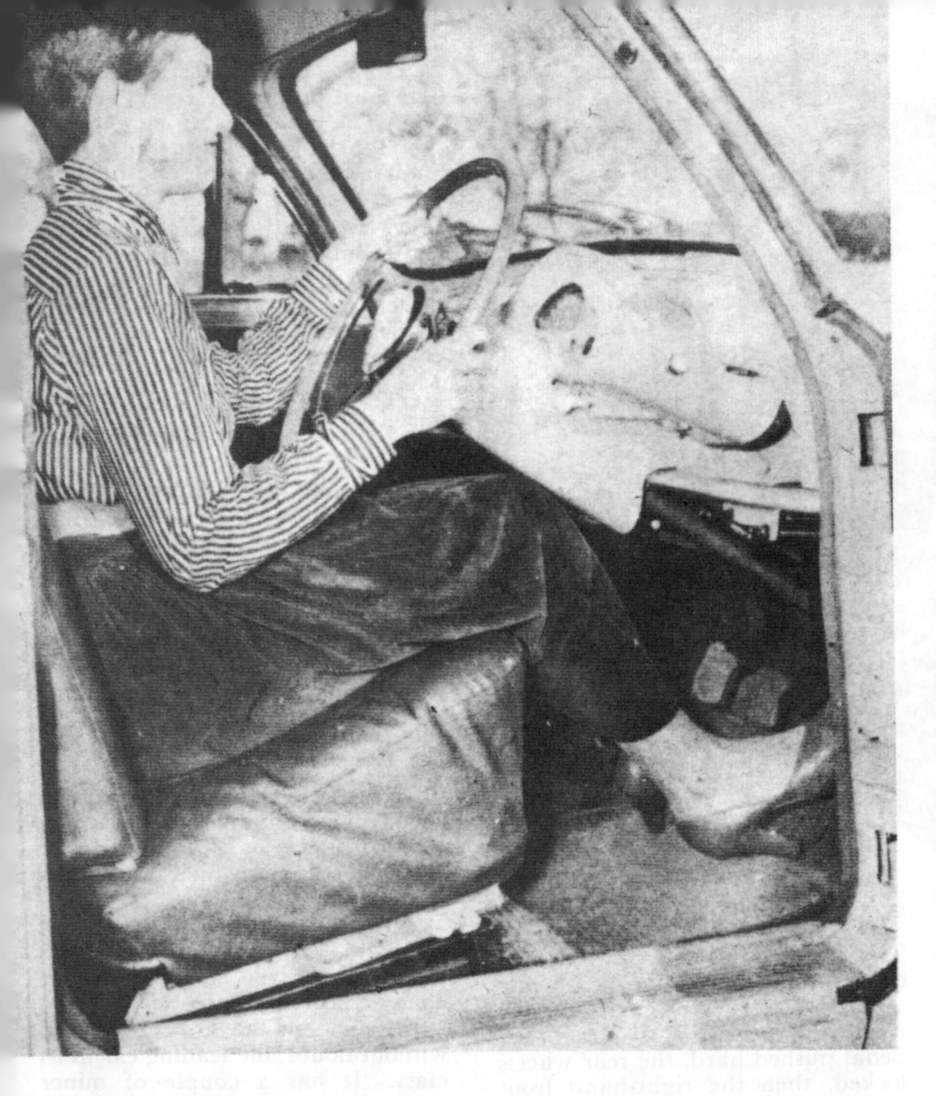

● **ABOVE: EASY access to the rear seats is provided by wide doors fitted with arm rests. Note the unusual door catches. LEFT: Driving position in the Zephyr is good for small people, but bigger ones would not be so happy. Pedals are the pendulum type.**

Zephyr seemed happiest doing about 17 m.p.h in first and 35 in second. Bottom gear was not equipped with synchromesh and was not easy to engage while the car was still on the move. This was no great hardship as second gear was able to handle, smoothly, any work from two m.p.h. upwards.

The gear change itself was smooth and firm and the synchromesh was impossible to beat with fast changes. Every time first or reverse gears were engaged, it was done with crashings of protest from the gearbox. It seems likely that the hydraulically-boosted clutch may have been out of adjustment.

Reverse gear in the Zephyr had a very low ratio. So low in fact, that the slightest heavy-footed operation produced a loud burst of wheelspin on dry concrete.

In first gear, too, wheelspin was easy to invoke on dry surfaces, and on gravel the wheels spun all the way to 50 m.p.h. when the car was really pushed.

In spite of this, there was no loss of directional stability on loose roads.

The Zephyr was very roadworthy in all conditions. On dirt roads the handling was nothing short of excellent. Belted along a corrugated gravel road at 80 m.p.h. the car held on very well, with no tail hop at all.

In the steering department there were both good and bad points. For normal running around town the Ford was a disappointment, thanks to the steering lock — the whole 41½ feet of it.

The steering was, however, light and accurate all the time, requiring little effort from the driver to swing around corners.

On narrow country roads the steering lock was nothing short of embarrassing, but as soon as the driver hits upon using the wheelspin available in first gear, to slide the tail around, the manoeuvre of U-turning becomes simple. Please note that the tail slide turn only applies to dirt roads.

A ground clearance of just over seven inches makes the Zephyr a good country car. A sump protector was also fitted.

Riding comfort in the car was good, even on the roughest surfaces. The suspension was soft — much softer than one would expect from an English car. It gave good road holding and plenty of comfort. There was no floating or sway at high speeds.

Some body drumming occurred on coarse surfaced roads, but never in disconcerting quantities.

Driven over rough cattle tracks, the front springs refused to bottom, although one of the rear semi-elliptics went to the limit in a huge water-filled hole, which would probably have wrecked a lesser car.

Inside the passenger compartment all the fittings were straightforward and not elaborate. Plastic roof lining looked to be a good feature and the seats were covered with artificial leather.

Armrests, fitted to the doors, proved comfortable on a long trip which is more than can be said for the seats. A person, with bigger specifications than average, such as myself, would find that the driving position lacked support.

Passengers, regardless of size, would find the car comfortable enough because they could move around on the seat. Leg room in the back was restricted slightly when the front seat was back as far as the adjustment allowed.

From the driving position, all round visibility was good. Through the windscreen the operator could see both the front mudguards through the one-piece curved windshield. Rearward vision was quite

● **LEFT: UNDER Zephyr's bonnet, most points are easily accessible. The battery is up forward on the right side of engine compartment and coil down on the left.**

good, too, although there was a blind spot between the rear window and the back door window. It was not a serious one.

The instrument panel was placed in front of the driver and the big, easy-to read speedo was accurate. A parcel shelf extending the full width of the car was deep and big enough to be called additional luggage space.

Only light pressure was needed to operate the pendulum-type clutch pedal.

In the stopping department, the brakes were light to use, but on concrete it was not possible to make the front wheels lock. Every braking test made the rear wheels lock and front ones just refused to co-operate. The car stopped in a straight line every time.

It was a different story on dirt roads, however. With the brake pedal pushed hard, the rear wheels locked, then the right-hand front one, causing the car to slew around sideways. A little careful adjustment would no doubt cure these bad habits.

The pistol-grip hand-brake was unusually good by modern car standards. From 30 mph on concrete it easily locked the rear wheels. A more progressive application slowed the car rapidly and without fuss.

No matter what kind of emergency, the Zephyr's steering was always up to the situation. Tail slides were always 100 per cent. controllable, and there were plenty of them when the car was being driven over loose-surfaced twisting roads at high speeds, because of the comparatively light load on the rear wheels.

The steering wheel had a plastic rim and spring-wire spokes with horn ring on the lower half of the helm. The horn, by the way, creates a shattering blast, quite in keeping with the character of the Zephyr.

Luggage space, in true American fashion, was very big, but unlike some of its Detroit relatives, the Zephyr's appearance didn't suffer for it. The spare wheel was mounted vertically on the right-hand side of the boot giving the luggage unobstructed room.

The headlights allowed a cruising speed of 55 mph at night, which was some 15 or 20 mph slower than the speed at which the car can be cruised in daylight.

Fuel consumption for the test Zephyr — driven really hard — was 23½ mpg. Driven normally owners could expect around 28 mpg.

To sum up, the Ford Zephyr is without doubt, the leading car in its class. It has a couple of minor faults, but its advantages certainly outweigh these a hundred to one.

It's fast and reliable. The oversquare motor seems assured of a long and prosperous life with many hours of pleasure to the owner. ●

SPECIFICATIONS

Engine: Six cylinders. Bore, 79.37 m.m.; stroke, 76.2 m.m. Capacity, 2262 c.c.
Pushrod operated overhead valves.

Engine: Six cylinders. Bore, 79.37 m.m.; stroke, 76.2 m.m. Capacity, 2262 c.c.
Pushrod operated overhead valves.
Compression ratio: 6.8:1.
68 b.h.p. at 4200 r.p.m.
16.15 m.p.h. per thousand revs in top gear.
M.p.h. at 2500 feet per minute piston speed in top gear, 80.78.

DIMENSIONS

Wheelbase, eight feet eight inches.
Front track, four feet two inches.
Rear track, four feet one inch.
Kerb weight, 23 cwt.
Price: £1133, incl. tax.

PERFORMANCE

Maximum speeds in gears: top, 80 m.p.h.; second, 50 m.p.h.; first, 30 m.p.h. Acceleration: Standing quarter mile, 22.3 seconds; 0-30 m.p.h., 5.2 seconds; 0-50 m.p.h., 12.1 seconds; 0-70 m.p.h., 30.2 seconds; 20-40 m.p.h. in top gear, 9.5 seconds; 20-40 m.p.h. in second gear, 5.5 seconds; 10-30 m.p.h. in second gear 4.4 seconds.

● **LUGGAGE space is good, and boot shape practical for family use.**

Used cars on the road

90

1952 FORD ZEPHYR SALOON

Price new .. £532 0s 0d
Purchase tax .. £297 1s 1d
Price secondhand £595 0s 0d

Acceleration from rest through gears:
to 30 *m.p.h.* .. 5·5 *sec*
to 50 *m.p.h.* .. 12·2 *sec*
to 60 *m.p.h.* .. 18·6 *sec*
20-40 *m.p.h. (top gear)* 8·4 *sec*
30-50 *m.p.h. (top gear)* 8·2 *sec*
Petrol consumption: 23-25 *m.p.g.*
Oil consumption: 1,200 *m.p.g.*
Speedometer reading: 26,545
Car first registered: December, 1952

The light stone colour of the Zephyr is enhanced by the use of a darker shade on the lower part of the body. The radio aerial is the only outwardly visible accessory. The turn-out was excellent from careful cleaning of the body to the polished tyres

A REMARKABLE success was scored by the Ford Zephyr and its smaller-engined brother, the Consul, from the moment of their introduction. These cars have now been in use for some years, and it is interesting to find out how a Zephyr has stood up to time and mileage. (A Consul used car test was published on October 8, 1954.) The car tested was provided by Gatehouse Motors, Ltd., 1, Hampstead Lane, Highgate Village, London, N.6. It had covered more than 26,000 miles and was a little over two and a half years old. This, of course, is not a great mileage or age for a well-produced, 2¼-litre, six-cylinder car, but it is enough to permit any inherent faults to reveal themselves.

There were two outstanding impressions when the car was put to the test. One was the smoothness of the engine, and the other the complete freedom from rattles—the body felt absolutely taut.

This car was in a light stone colour, with the addition of a darker band round the lower part. All the synthetic baked enamel paintwork was in good, undamaged condition. There were no lighter or darker panels suggesting previous accident damage. The chromium plating was generally good, but there were signs of slight rust on some parts, including the window surrounds.

Hide upholstery is used, and the matching red carpets were in almost unworn condition. The radio is fitted neatly in the centre of the parcels shelf, with the heater-demister controls below it

The interior was upholstered in red leather (with leather-cloth on the non-wearing surfaces), and this, too, was in good condition, except for some scratches on the front, bench-type seat. It was evident that the red carpets had been kept covered, for they were virtually unworn. A touch of luxury, in addition to that given by the hide upholstery, was added to the interior by the built-in radio, which worked very well, and by the heater-demister system, also built-in and effective.

The engine smoothness was noticed as soon as the car was started—accomplished at a touch on all occasions. Acceleration was particularly good. The model has a reputation for accelerative ability, and the car tested proved to gain speed even more quickly than the car that was subjected to a full road test in November, 1951. This may well have been because the extra mileage had made the engine more free, enabling it to give more power.

This acceleration was exhilarating from a family car, and was backed up by the efficiency of other components that add to roadworthiness. There was little play in the steering, which was light to operate. The brakes were entirely up to the requirements of normal fast driving and did not pull to either side under heavy pedal pressure. The suspension gave a comfortable ride and felt safe. Absence of pitching showed that the spring dampers were in sound condition.

The main fault was a vibration period at about 55 m.p.h. on the almost accurate speedometer. At this speed there was a very slight steering shake, made to appear more formidable by a coincident transmission tremor. However, the vibration as a whole existed only within the span of a few m.p.h. and the acceleration was such that the period could be passed through quickly. The car would cruise very comfortably in the sixties, and pass the true 70 mark without any sign of distress. The gears were occasionally rather stiff in the change. Oil consumption was slight in normal driving, rising, as was to be expected, when high speeds were maintained. The approximate consumption figure quoted in the data table is an estimate that should apply when the car is made to work quite hard.

There were four tyres in good condition, three of them Indias. The spare was an unused Firestone, presumably remaining from the original equipment. The fifth was well worn. All the controls and instruments worked efficiently, the latter being limited to a speedometer with total mileage recorder only, a fuel gauge, an ammeter, an oil pressure warning light, and a main beam indicator.

This Zephyr was gratifyingly powerful, and it felt comfortable, well built, and in generally very sound condition.

A CAR LIFE *Report for the Auto Buyer*

ZODIAC:

The lines of the little car are smooth and it blends well into the U. S. motoring scene.

By G. M. LIGHTOWLER

COMPETITION is the life of trade, says the business axiom, and so strongly does the Ford Motor Company believe in this principle that it is, in a sense, competing with itself by spotting dealers for its British cars all over the country. True, there is a considerable difference in the price brackets between the American product and its British cousins—Zodiac, Zephyr, Consul, Anglia and Prefect—but the latter models are definitely competing for the U. S. auto business in the low-price field.

This report will deal with the Zodiac, one of the most interesting low-priced foreign sedans on the American market, and basically the same car as the Ford Zephyr, the model that won the 1953 Monte Carlo Rally. We were fortunate enough to be on hand to drive the winning car both before and after its victorious trial. To say that we were impressed by the Zephyr is an understatement and Maurice Gatsonides, who drove the car, was more than delighted

Modest use of chrome adds to Zodiac's appearance achieved by the stylist and engineer.

Luxury Car in Miniature

This English-made Ford product is a slicked-up version of the Zephyr—a small car by American standards but an efficient piece of machinery

with its performance and reliability. The Zephyr win went down in history as a victory for the lower-priced sedans, the Zephyr being the cheapest car ever to win the classic Monte.

Today's Zodiac is the luxury version of today's Zephyr. It is a complete automobile and the price paid for it ($2095) includes all necessary accessories. The impressive list of these items, which often jack up the price of a car, includes: back-up light, two fender mirrors, turn signals, windshield washers, white sidewall tires, chrome rim embellishers, two-tone paintwork, locking gas-tank cover, oil filter, oil-bath air filter, hydraulic clutch, padded tool box, tools, two-tone leather upholstery, electric clock and cigarette lighter. The only two items which are not included as standard equipment are a heater ($57) and a radio. Despite its smallness in relation to the conventional sedan, the Zodiac blends easily into the American scene. Although inclined to be boxlike in contour, the lines of the car are smooth and dignified with a front-end treatment which reminds one of a well-known expensive sports car. It is a luxury car in miniature.

The change of color of the two-tone paint job is made at the chrome line which delicately cuts the body. There are three standard color combinations: grey and green, grey and blue, and grey and fawn. The car tested was finished in the grey-and-blue combination and drew complimentary remarks from all who saw it.

Two-tone leather of good quality on the seats matches the color treatment on outside of auto. Rear seat accommodates three and has drop arm-rest for two.

The absence of large areas of chrome is gratifying; where chrome is used it is handled with moderation and subtility. Although the European manufacturer is turning his attention to the wraparound windshield, it is hoped, after viewing the Zodiac both from outside and inside, that the Ford Motor Company of Dagenham will not change the excellent curved windshield now used. All in all, the Zodiac is distinctive without being ostentatious.

Opening the doors and inspecting the rubber sealing and general construction of the body, one sees a high quality of craftsmanship. A similar standard of engineering acumen is displayed in the engine compartment. The placing of the engine accessories shows considerable planning, and was evidently studied from the view of both operation and maintenance. There are no loose ends lying about, or small items carelessly screwed into places on which they apparently just happened to fall—as we have observed in several cars tested recently.

From behind the steering wheel, one is immediately impressed by the luxury of appointments, but there are certain exceptions to the general fine craftsmanship. The aircraft-type instrument panel is much below par. The figures on the panel are neither distinct nor well finished; the lights which indicate a failing generator supply, or a shortage of oil, are shoddy, consisting as they do of crude translucent crosses formed by the open tops of Phillips-type screws inserted into the panel. The gas-and-oil pressure gauges are small and poorly graduated; the shield which surrounds the upper part of the steering column and the top joint of the gearshift linkage is unfinished, and the unsightly gap in the casing where the gear lever enters is quite unnecessary.

Why the British manufacturer falls down on items like these is frankly beyond comprehension, for the remainder of the car's interior appointments are of a high quality.

The turn-signal switch is situated on

Trunk is small by American standards. License light also illuminates trunk.

the center of the steering wheel, and the column gear-shift lever to the right in readiness for the quickest flick of the wrist. The steering wheel itself is positioned at a comfortable angle which encourages correct placing of the hands in the lower segment. All control knobs and switches are accessible without undue stretching or interference.

The positioning of the brake and accelerator is excellent, for toe-and-heel double-clutching can be executed without difficulty. The foot need not be lifted off the accelerator to apply the brake, but merely pivoted to the left. The provision of a rest to the right of the accelerator is a boon to the traveler on long journeys when he can maintain a constant speed. The headlight dip switch is large and placed well over to the left, leaving ample room for the left foot to rest against the firewall when it is not actually operating the clutch.

The view from the driver's seat could scarcely be better, with both front fenders being clearly visible and greatly assisting the accurate placing of the automobile. All-around visibility is excellent and the blind spots that exist offer no problem after a few miles at the wheel. The two mirrors attached to the front fenders, coupled with the interior mirror, give adequate indication of anything approaching from the rear.

The interior of the automobile is generally pleasing with the two-tone leather upholstery and pile carpeting. The provision of an open parcel tray which stretches the full width of the car beneath the instrument panel is a feature which could well be copied by other manufacturers. The placing of the electric clock in the roof between the two sun visors is ingenious. The passenger's sun visor has a mirror fitted on its back.

If, upon looking at the Zodiac for the first time one gets the impression that there is limited room inside, a short journey either in the front passenger's seat or in the comfortable rear compartment will soon dispel the idea. There is room for three in the rear seat, which has a center let-down arm-rest for use with only two passengers. The front bench-type seat, due to its height and incline, provides good leg support and although we are consistent proponents of the individual bucket-type seats, the bench seat in the Zodiac is of such construction that adjustments do not interfere with the comfort of the passenger who has ample room to stretch his legs.

The luggage trunk by American standards is small and demands careful selection of items for a lengthy trip. The contents of the padded toolbox, which is situated to the right of the spare wheel and tire, calls for commendation. There are sufficient well-made wrenches, screwdrivers and the like to carry out any emergency repair at the roadside, and a substantial jack is also provided.

The handling characteristics of the Zodiac are similar to those of a sports car and it is a dream to steer it around a corner with the assurance that it is going to stay firmly on its four wheels. The steering is delicate, yet positive. There is a pleasant sensation of slight understeer, and drifting a corner is considerably more simple an operation than with some of the recognized sports machines. However hard we tried to promote it, there was no objectionable tire squeal.

These qualities of roadability were admirably demonstrated last year when a Ford Zephyr (the machine basically identical to the Zodiac) finished fourth in the Linden, New Jersey, 12 Hours Race, being beaten only by sports cars of greater capacity and a much higher price bracket. The tire wear on the Zephyr in this race was far less than any of the other cars and pit stops were made for only one change of tires and gas and water.

This road-holding is in keeping with the vivid performance of the engine. This is no underpowered foreign job—a slight pressure of the accelerator will prove this. Getaway will surprise even the hardened driver who is used to commanding 200 bhp. The torque should satisfy any American, for the engine is as happy in top gear at 10 mph. as it is at 80.

The flexibility of the engine and gearbox, coupled with the excellent road-holding, makes the Zodiac as much a tight city car as a speedy transport on the open highway. Parking in a congested area is simple and needs no assistance from power equipment.

The Zodiac is powered by a 6-cylinder ohv. engine of 2262 cc. capacity (138 cubic inches) with a bore and stroke of 79.37 x 76.20 mm. When turning over at 4200 rpm. the engine

Six-cylinder engine develops 71 bhp., takes car to speeds exceeding 80 mph.

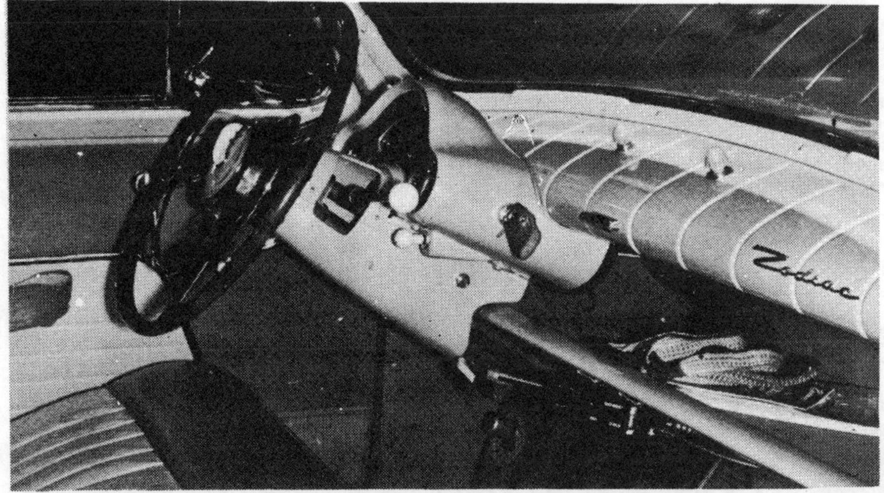

Parcel shelf stretches full width of car. Hole in casing around upper part of steering column and gear lever linkage detracts from appearance of interior.

The Zodiac is docile in traffic and potent on the road

develops 71 bhp., while at 2000 rpm. the torque register reads 112 foot pounds. The detachable cylinder head is of cast-iron, and has wedge-shaped combustion chambers that give controlled burning of the fuel and air mixture. The overhead valves work in conjunction with push rods operated from harmonic cams in the cylinder block.

Engine lubrication is pressure-fed by a submerged gear-type pump; a full-flow type oil filter is fitted directly to the block.

The fuel system utilizes a single down-draft carburetor which has a special choke control and is fitted with a combined air cleaner and silencer. The fuel tank at the rear holds 10.8 U. S. gallons.

Power from the engine is transmitted to the rear by way of a single dry-plate clutch whose operation is hydraulically assisted from pedal to heavy duty ball-bearing thrust release. The three-speed gear-box has synchromesh on top and second gears. The gears are helically cut.

The improved rear axle has a ¾-floating hypoid final drive. The open drive shaft has heavy duty roller and ball bearings, and the universal joints have needle roller bearings.

Spare parts and service are still, in many cases, a headache for the owner of an imported car, but this does not apply to Ford's English cars which are made by the Ford Company of Dagenham, England. The marketing and servicing of all the English Ford models are the responsibility of the Ford Company at Dearborn which maintains the spare parts depots. Every major city in the United States has its English Ford dealer who carries basic supplies in stock. Furthermore, the U. S. Ford dealers can obtain parts for the English Fords through Dearborn in a matter of hours. Another asset is the fact that Dagenham has standardized its nuts and bolts to conform to American measurements.

Summarizing, we would say that the Zodiac fulfills the requirements of the American who wants a moderate-priced second car. It has European handling characteristics, yet a body style that fits into the American picture. It is docile for traffic driving, yet potent on the open road. It offers class for a modest outlay.

To the person who is sceptical about imported makes of cars, we would say, "Drive a Zodiac or Zephyr and see for yourself." Whatever conclusions you come to, you will at any rate be able to say that you have driven a replica of the car that won the 1953 Monte Carlo Rally. ☆☆

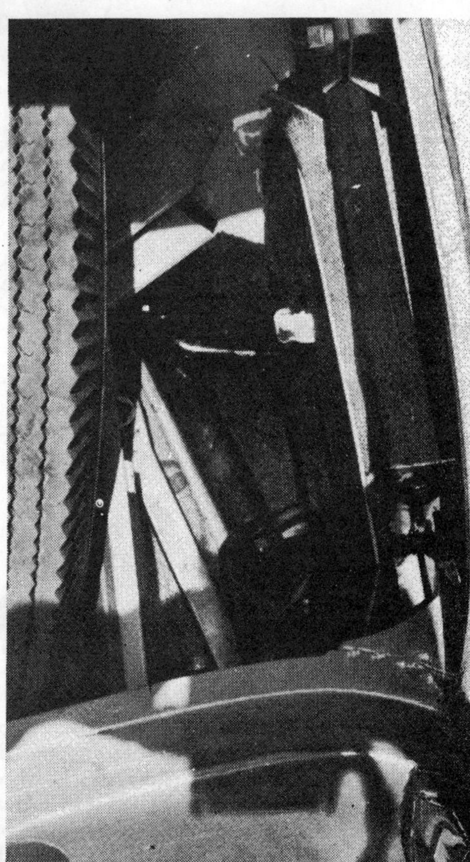

Toolbox to right of spare wheel is felt lined, has good tool selection.

Sun visor at right has a mirror on the reverse surface. The electric clock in roof can be seen by passengers in rear.

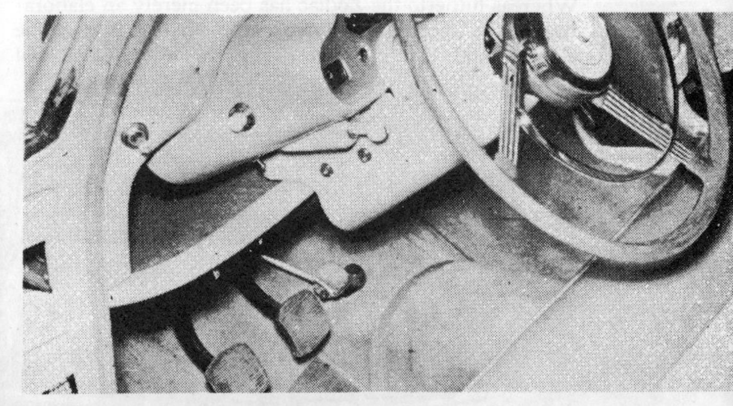

Positioning of the pedals is remarkably good, encouraging use of the toe-and-heel double-clutching technique of driving.

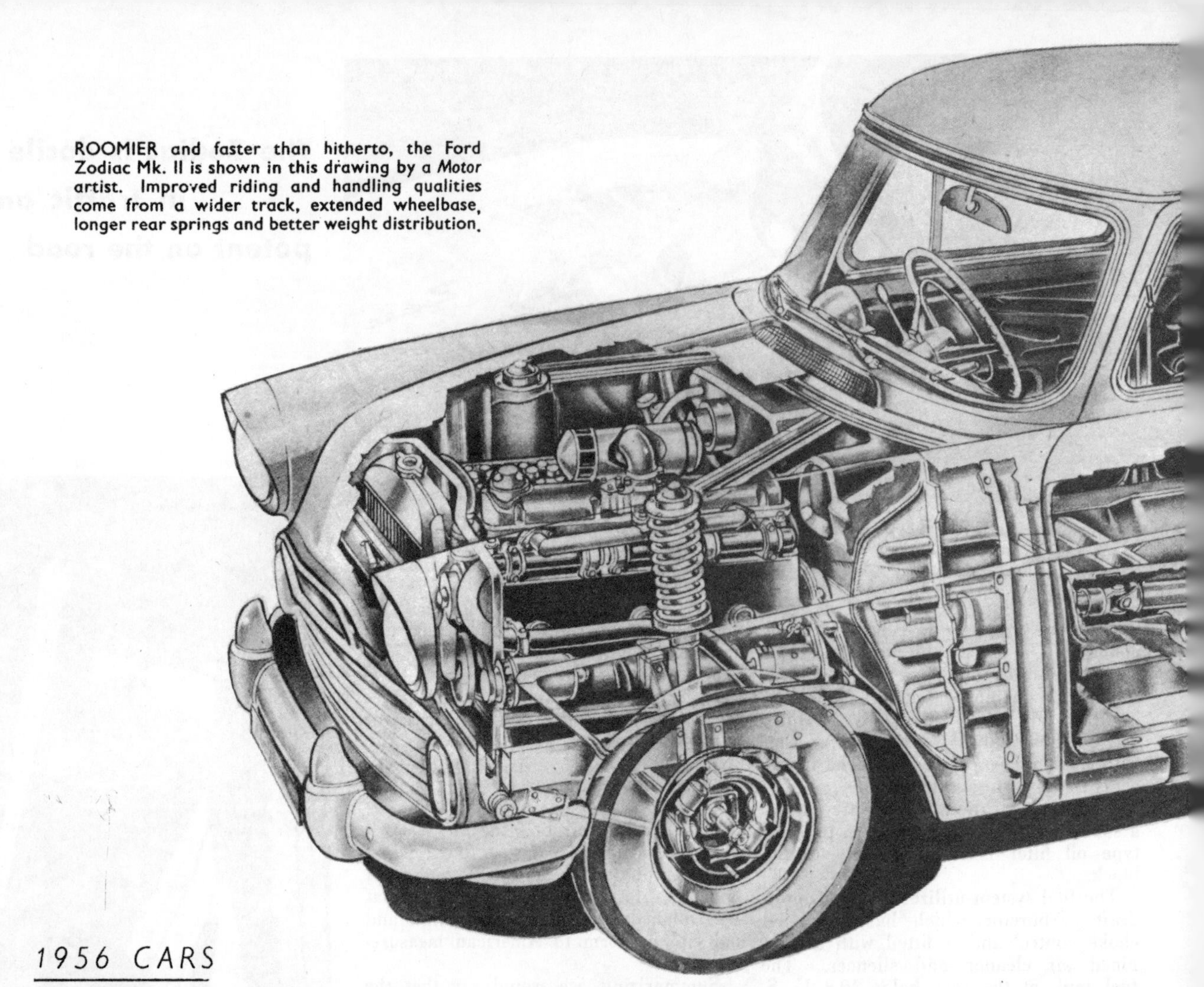

ROOMIER and faster than hitherto, the Ford Zodiac Mk. II is shown in this drawing by a *Motor* artist. Improved riding and handling qualities come from a wider track, extended wheelbase, longer rear springs and better weight distribution.

1956 CARS

FORD CONSUL, ZEPHYR and ZODIAC

BIGGER and better than ever is the natural description of the Mk. II Consul, Zephyr and Zodiac models which have been announced by the Ford Motor Co. of Dagenham. Increases in wheelbase, track and overall dimensions are matched by increases in cylinder bore and stroke, and higher gear ratios promise more effortlessly economical fast cruising than hitherto. The new cars are closely similar in layout to their predecessors, so that full advantage can be taken of lessons learned since the Consul and Zephyr first appeared in October, 1950, but almost every component part of them has been re-designed. Whereas hitherto the Zodiac has been merely an elaboration of the Zephyr, it is now a longer car overall than the Zephyr and has its own distinctive frontal and tail treatment and wing pressings as well as dual colours.

* * *

At the heart of the new cars, larger successors now appear to the 4- and 6-cylinder engines which did so much to popularize "over square" cylinder dimensions. As hitherto, the 4-cylinder Consul has the same bore and stroke as the 6-cylinder Zephyr and Zodiac, but these dimensions are now 82.55 mm. by 79.5 mm. instead of 79.37 mm. by 76.2 mm. With the bore and stroke both increased by approximately 4%, swept volumes go up by 13%, to new figures of 1,702 c.c. for the Consul and 2,553 c.c. for the Zephyr and Zodiac.

Simultaneously with these increases in size, the decision has been taken to standardize a compression ratio high

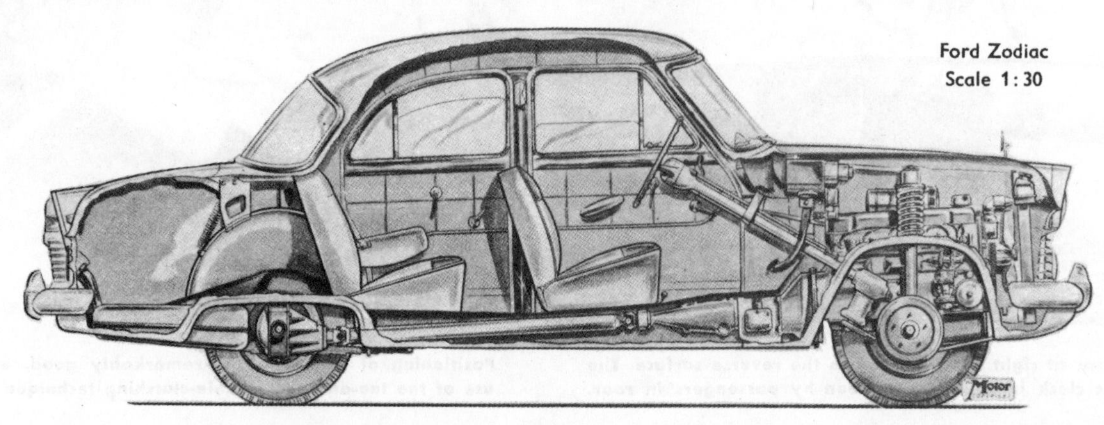

Ford Zodiac
Scale 1 : 30

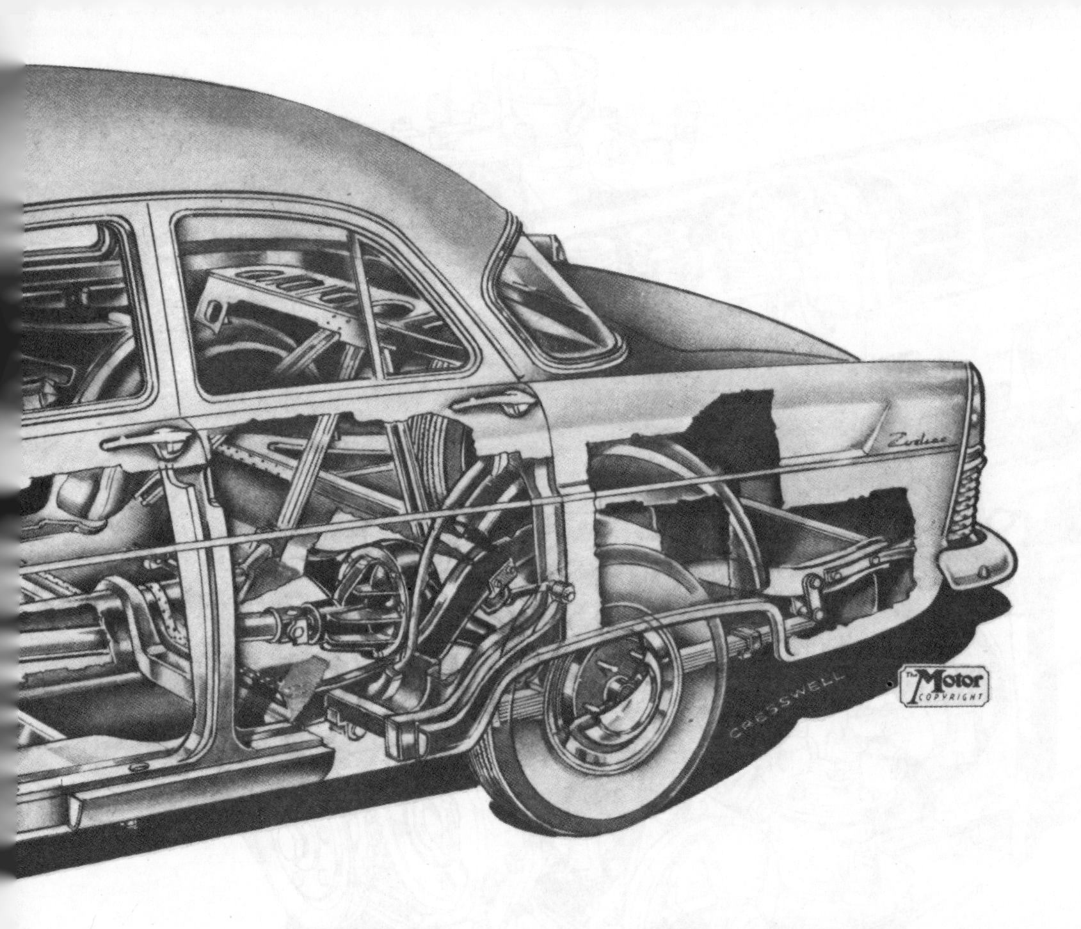

Enlarged Engines and Bodywork for Mark II Versions of Three Popular Cars

RE-DESIGNED

enough to take full advantage of premium-grade petrol. It will still be possible to obtain either engine with a compression ratio of 6.9/1 if desired, but normal deliveries will be of cars with a 7.8/1 compression ratio which provides approximately 6% more torque in the middle of the speed range and 7% more power at peak r.p.m. Thus, the Consul Mk. II engine now develops 59 b.h.p. whereas its predecessor gave 47 b.h.p., and the Zephyr-Zodiac Mk. II engine gives 86 b.h.p. as against the 68 b.h.p. and 71 b.h.p. of its forerunners. Although the new cars are longer and wider than hitherto, they are stated to have almost unchanged weights, so that substantial gains in overall performance can be expected.

Advantage has been taken of the more powerful engines to use higher top gear ratios on all cars, 4.11/1 instead of 4.556/1 on the Consul and 3.90/1 instead of 4.444/1 on the Zephyr, with in each case the same wheel and tyre sizes as hitherto. Quieter cruising at high speeds is given by the new gearing, which should also ensure that under many driving conditions fuel economy is not impaired by the increases in engine size.

Conspicuously, the new cars have completely fresh bodywork designs. The main passenger-carrying steel shell is the same for all three, but the distance from the windscreen to the front suspension is greater on the 6-cylinder cars than on the Consul, and the Zodiac gains in appearance by having more rear overhang and overall length than the Zephyr. The new body dimensions allow generous head and legroom at the back as well as at the front, and extra breadth makes them genuine six seaters. Points which show the results of experience are that the major components of the latest cars have been re-arranged so that an appreciably greater proportion of the weight is carried on the driven rear wheels, that both cars now have

Ford Consul
Scale 1:30

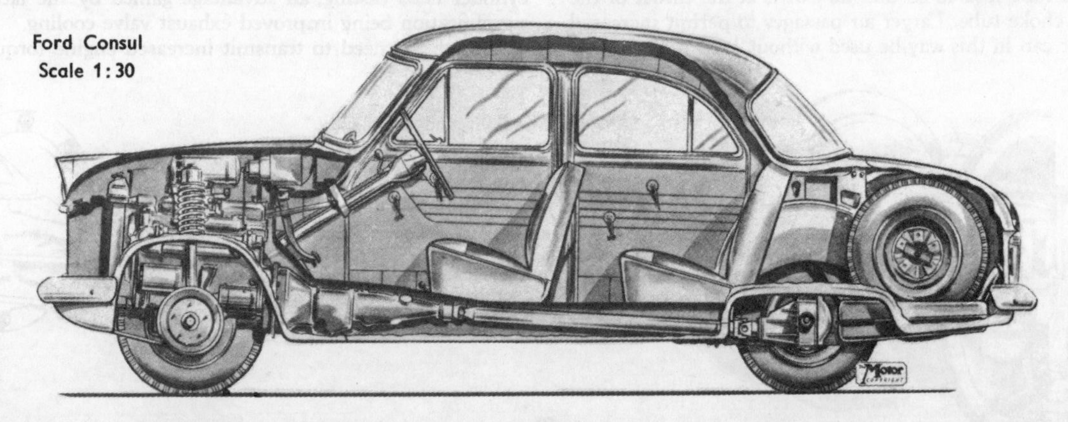

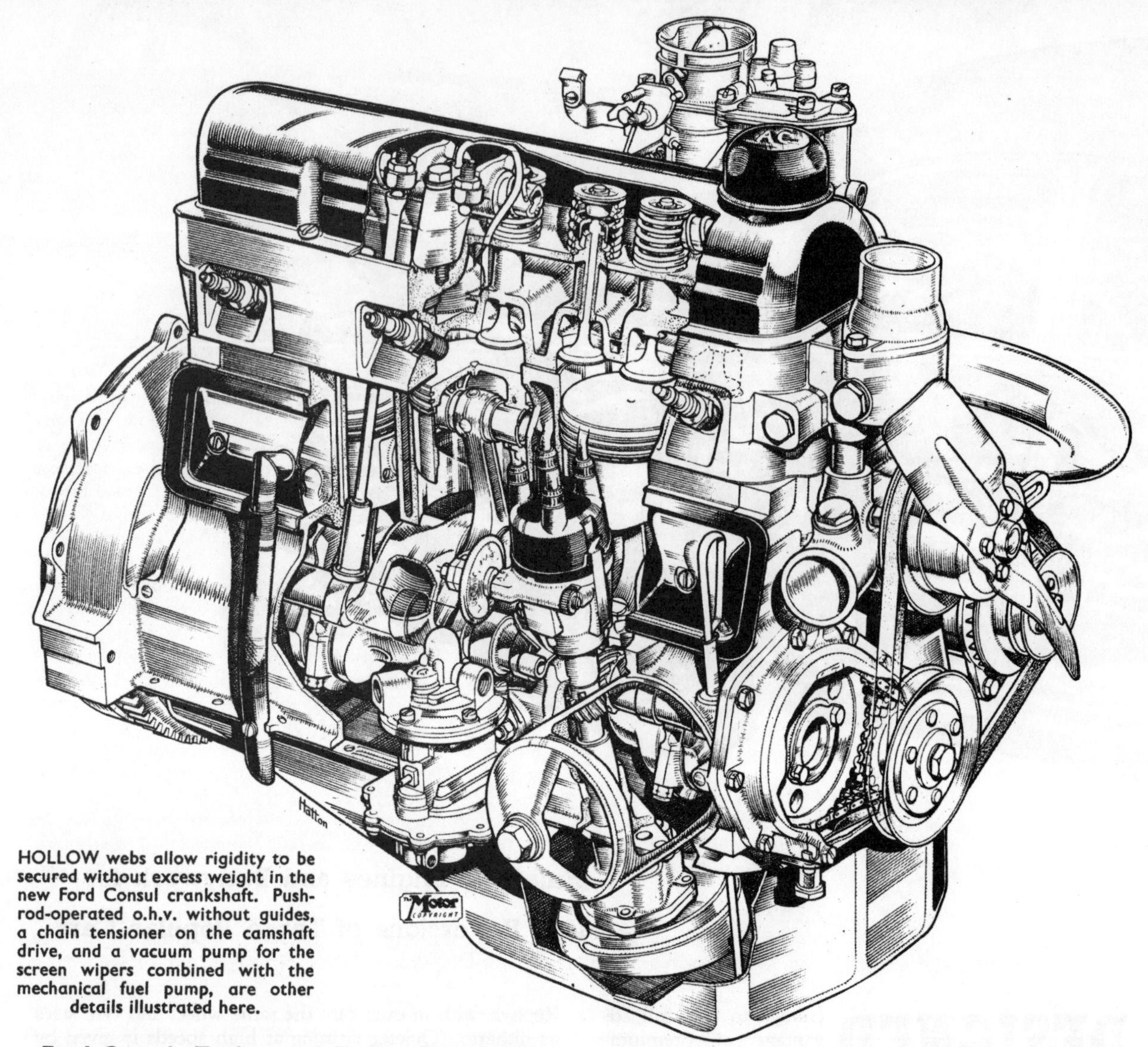

HOLLOW webs allow rigidity to be secured without excess weight in the new Ford Consul crankshaft. Push-rod-operated o.h.v. without guides, a chain tensioner on the camshaft drive, and a vacuum pump for the screen wipers combined with the mechanical fuel pump, are other details illustrated here.

considerably more compact turning circles, and that brake lining areas and fuel-tank capacities have been increased more than proportionately with the engine enlargements.

In appearance the new 4- and 6-cylinder engines closely resemble their predecessors, which have established an enviable reputation for durability. The unusual exhaust manifold, consisting simply of a straight pipe with openings in one side clamped to the cylinder head, continues to be used. There is, however, a new type of Zenith downdraught carburetter, used on both engines with appropriate choke sizes, which is conspicuously lower than the familiar design. This carburetter, which is described elsewhere in this issue of *The Motor*, is of the "boost venturi" type, with its jets discharging fuel into a miniature choke tube located so that its exit is at the throat of the main choke tube. Larger air passages to permit increased power can in this way be used without loss of flexibility.

Inside the engines, the most interesting innovation is that new cast crankshafts giving larger bearing areas as well as a longer stroke have hollow webs between big end journals which (apart from the usual lubricant drillings) are solid. This design allows rigidity to be combined with low weight. Inclined in-line overhead valves continue to be operated by pushrods from a low-set crankshaft, the 4-cylinder engine alone requiring a spring loaded tensioner for its camshaft driving chain. Details which typify the Ford policy of simplification, using the minimum number of parts which tests show to be necessary for good results, are the use of bolts instead of studs and nuts for securing the cylinder head to the block, and the elimination of separate valve guides inserted into the cylinder head casting, an advantage gained by the latter simplification being improved exhaust valve cooling.

Despite the need to transmit increased engine torque,

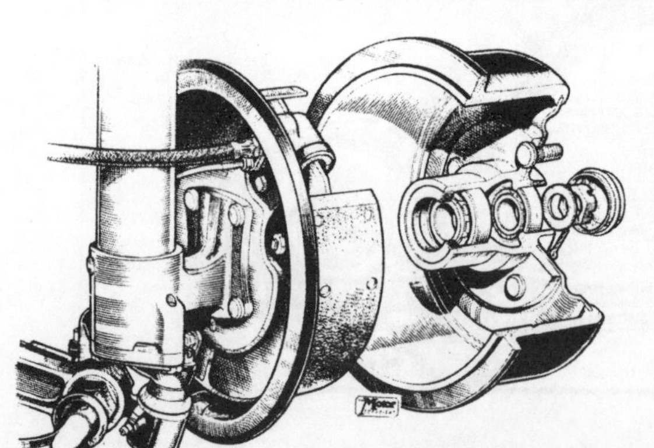

SAVING of weight and complexity results from combining one universal joint yoke with the rear axle pinion shaft (*right*).

GAIN in braking area comes from front drums 0.5 inch wider than hitherto, the rubbing surfaces being of iron cast onto a pressed steel flange (*left*).

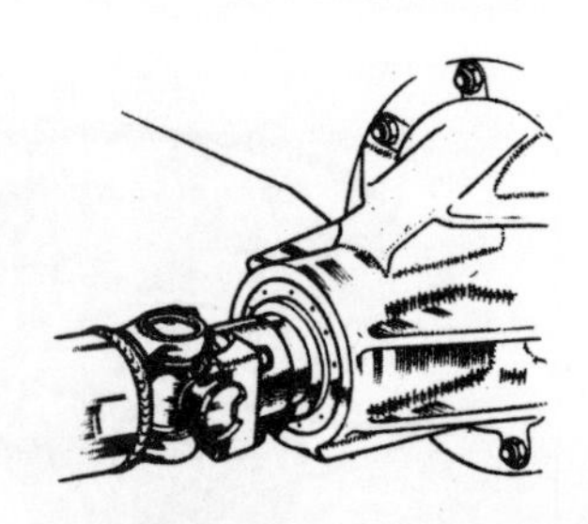

INDIVIDUAL radiator air intake grilles and wing pressings distinguish the three models, covering a range of prices from the luxurious Zodiac (*above, right*) via the similarly-engined Zephyr (*above*) to the 4-cylinder Consul (*below*).

the new hydraulically-operated clutches disengage with lighter pedal pressures than hitherto, thanks to the use of adjustable over-centre helper springs on the hanging pedals. Three-speed gearboxes continue to be used, with long rear extensions to reduce the length of open propeller shaft. The telescopic splines of the propeller shaft are enclosed in the tail of the gearbox, and the usual flanged joint between propeller shaft and hypoid rear axle has been eliminated by making one universal joint yoke integral with the pinion shaft. All models are available with the Borg-Warner semi-automatic overdrive as an optional extra, 6-cylinder cars retaining their rear axle ratio unchanged when this is fitted but the Consul having its final drive altered to give a livelier direct top of 4.444/1 and an effortless overdrive top of 3.11/1 instead of the usual 4.11/1 ratio.

Improved riding comfort and road-holding have been obtained on the new cars, without departure from the fundamental principles of their shorter and narrower predecessors. I.F.S. continues to be by telescopic legs which combine coil springs and hydraulic shock absorbers, supplemented by single transverse links and brake reaction stays which are combined with an anti-roll torsion bar; but the design has been re-worked for better performance and life. Longer semi-elliptic rear springs have the axle clamped to them forward of their centre, location of the rubber bump stops above the springs some inches forward of the axle saving height and giving more gently progressive action.

To simplify the world-wide stocking of a wearing part, the tyres (and wheels) of the new cars have been kept the same as those of their predecessors, although the new bodies allow room for oversize tyres on all models. A

FORD DATA

	Consul Mk. II	Zephyr and Zodiac Mk. II
Engine dimensions		
Cylinders	4	6
Bore	82.55 mm.	82.55 mm.
Stroke	79.5 mm.	79.5 mm.
Cubic capacity ...	1,702 c.c.	2,553 c.c.
Piston area	33.16 sq. in.	49.74 sq. in.
Valves	Pushrod o.h.v.	Pushrod o.h.v.
Compression ratio ...	7.8 (optional 6.9)	7.8 (optional 6.9)
Engine performance (7.8 compression ratio)		
Max. power	59 b.h.p.	86 b.h.p.
at	4,200 r.p.m.	4,200 r.p.m.
Max. b.m.e.p.	134 lb./sq. in.	132 lb./sq. in.
at	2,300 r.p.m.	2,000 r.p.m.
B.H.P. per sq. in. piston area	1.78	1.73
Piston speed at max. power	2,190 ft./min. (With low c.r., 55 b.h.p. 126 lb./sq. in. b.m.e.p., 1.66 b.h.p./sq. in. piston area)	2,190 ft./min. (With low c.r., 80 b.h.p. 124 lb./sq. in. b.m.e.p. 1.61 b.h.p./sq. in. piston. area)
Engine details		
Carburetter	Zenith 34 WIA downdraught	Zenith 36 WIA downdraught
Ignition timing control	Centrifugal and vacuum	Centrifugal and vacuum
Plugs	14 mm. Champion	14 mm. Champion
Fuel pump	Mechanical, with vacuum pump	Mechanical, with vacuum pump
Fuel capacity	11 gallons	11 gallons
Oil filter	AC full-flow	AC full-flow
Oil capacity (including filter)	7¼ pints	8½ pints
Cooling system ...	Pump, fan and thermostat	Pump, fan and thermostat
Water capacity (excluding heater) ...	16½ pints	22½ pints
Electrical system ...	12-volt	12-volt
Battery capacity (20-hour rate)	45 amp.hr. (optional 57)	57 amp.hr. (optional 72)
Transmission		
Clutch	Single dry plate	Single dry plate
Gear ratios		
Top (s/m)	4.11	3.90
2nd (s/m)	6.75	6.40
1st	11.67	11.08
Transmission—contd.		
Rev.	15.86 (With optional overdrive, ratios are:—O/d top, 3.11, direct top, 4.44, o/d 2nd, 5.11, direct 2nd, 7.29, direct 1st, 12.61, reverse, 17.15.)	15.06 (With optional overdrive, ratios are:—O/d top, 2.73, direct top, 3.90, o/d 2nd, 4.48, direct 2nd, 6.40, direct 1st, 11.08, reverse, 15.06.)
Prop. shaft	Hardy Spicer open	Hardy Spicer open
Final drive	Hypoid bevel	Hypoid bevel
Chassis details		
Brakes	Girling hydraulic, 2 l.s. front	Girling hydraulic, 2 l.s. front
Brake drum diameter	9 in.	9 in.
Friction lining area ...	147 sq. in.	147 sq. in.
Suspension:		
Front	Coil spring i.f.s.	Coil spring i.f.s.
Rear	Semi-elliptic	Semi-elliptic
Shock absorbers:		
Front	Incorporated in i.f.s.	Incorporated in i.f.s.
Rear	Lever-arm hydraulic	Lever-arm hydraulic
Wheel type	Steel disc, 4J rim	Steel disc, 4½J rim
Tyre size	5.90—13	6.40—13
Steering gear	Worm and peg	Worm and peg
Dimensions		
Wheelbase	8 ft. 8½ in.	8 ft. 11 in.
Track: Front	4 ft. 5 in.	4 ft. 5 in.
Rear	4 ft. 4 in.	4 ft. 4 in.
Overall length ...	14 ft. 2 in. (without over-riders)	14 ft. 10½ in. (Zodiac, 15 ft. 0½ in.)
Overall width	5 ft. 7 in.	5 ft. 7 in.
Overall height (laden)	4 ft. 11½ in.	4 ft. 11¾ in.
Ground clearance ...	6½ in.	6¾ in.
Turning circle (between kerbs)	35 ft.	36 ft.
Dry weight	21 cwt.	22¾ cwt. (Zodiac 23¼ cwt.)
Performance factors (at dry weight)		
Piston area, sq. in. per ton	31.6	43.8 (Zodiac 42.8)
Brake lining area, sq.	140	129 (Zodiac 126)
Top gear m.p.h. per 1,000 r.p.m.	16.6	18.3
Top gear m.p.h. at 1,000 ft./min. piston speed...	31.8	35.1
Litres per ton-mile ...	2,930	3,680 (Zodiac 3,600)

DISTINCTIVE buttoned leather upholstery is used on the Zodiac, a drawing (*left*) also showing the courtesy light which has a lens to spotlight the ignition keyhole, and how a radio speaker can be mounted below the rear parcel shelf. Grouped instruments, hanging pedals and the glove locker and parcel shelf in front of the passenger are exemplified (*below*) by a Zephyr.

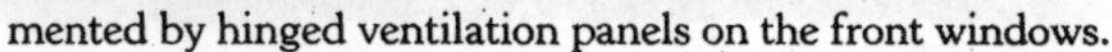

Ford Consul, Zephyr and Zodiac Redesigned - - - - Contd.

substantial increase in brake lining area has, however, been secured, by widening the front brakes from 1¾ in. to 2¼ in., weight transfer during braking making it desirable that these should dissipate a larger proportion of the energy than do the rear brakes. As a drawing shows, unsprung weight has been saved by casting a rigid iron drum onto a pressed steel flange, the front brakes continuing to be of two-leading-shoe pattern. A 12-volt battery is mounted to the right of the engine.

Completely new 4-door 4-light saloon bodies have been designed for these cars, of integral steel construction and needing no separate chassis frames. Internal widths of 56.5 inches at the front seat, 55.9 inches at the rear seat make these new bodies genuine six-seaters, and 6-foot tall men are no longer cramped for legroom or headroom in either front or back, another point being that the front floor is now level with the door openings instead of recessed below them. Neat new facia panel layouts are used on all models, with instruments facing the driver, a glove box and separate parcel shelf in front of the passenger.

Externally, all cars feature hooded headlamps incorporated in long wings, and use wrap-around curved glass windscreens and rear windows to provide good all-round visibility—the high rear wings as well as the front wings can be seen from the driving seat. Photographs show the distinctive frontal treatments given to the three models, and also the longer wings and metallic rear panel which add extra distinction to the two-colour Zodiac—external foglamps and mirrors are no longer depended upon to identify this model from the Zephyr. On all three cars, lift-up lids give access to usefully roomy luggage compartments; the spare wheel is stowed slightly away from one side, facilitating its loading or unloading and providing a secure place for small items between the wheel and the side of the locker. Petrol fillers are now concealed behind a rear number plate mounted on a spring-loaded hinge.

Integral provision is made for a neat Smiths fresh-air heating system, wide channels in the bonnet top panels leading air to an intake below the windscreen (provided with large drains for rainwater) which immediately adjoins the compactly grouped heat exchanger and circulating fan. Winding windows on all doors are supplemented by hinged ventilation panels on the front windows.

At Boreham airfield, we had the opportunity to drive an early production-line Zephyr and late prototypes of the Consul and Zodiac, before they were publicly shown. Comfortable riding, nicely balanced handling on fast corners despite patches of slushy snow on the airfield perimeter, and much improved turning circle were evident on all three, and the 6-cylinder cars showed a pleasing willingness to go up to speedometer readings of well over 60 m.p.h. in second gear, which should greatly facilitate overtaking in fast traffic. True maximum speeds in excess of 75 m.p.h. for the Consul and near to 90 m.p.h. for the Zephyr and Zodiac are suggested, while the latter, having substantially more weight on the driven rear wheels than had their predecessors as well as more power, are said to be capable of covering a standing ¼ mile in under 20 sec.

COACHWORK SURVEY (Manufacturer's figures)
FORD Consul Mk. II, Zephyr Mk. II and Zodiac Mk. II

Interior Length	Inches	Rear Seat—contd.	Inches
(a) Facia panel to rear squab	80	(k) Cushion depth, front to rear ...	18.3
Front seat		(l) Total length, (j) + (k)	28.1
(b) Overall width ...	56.5	(m) Seat height above floor	14.8
(c) Brake pedal to edge of seat ...	18.6	(n) Headroom above seat	34.5
(d) Cushion depth, front to rear ...	17.3	(o) Total height, (m) + (n)	49.3
(e) Total length, (c) + (d)	35.9	**Doors**	
(f) Seat height above floor	11.7	(p) Front door width, min./max. ...	30½/35
(g) Headroom above seat	35.5	(q) Rear door width, min./max. ...	22/29
(h) Total height, (f) + (g)	47.2	**Windscreen**	
Rear seat		(r) Width between pillars (max.) ...	53
(i) Min. width of seat	55.9	(s) Height at centre...	16.0
(j) Knee room	9.8		

NOVEL use is made of a metallic surface to give the distinction to the Zodiac tail. The petrol filler is concealed by a hinged cover upon which the number plate may be mounted, and accommodation for a large amount of luggage is provided in the locker at one side of which is mounted a spare wheel and tyre.

✓✓✓✓✓ MEANS TOP RATING	FORD ZEPHYR
PERFORMANCE ✓✓✓✓☐	One of the top performers in the lightweight imported car class. Acceleration from 0 to 60 mph can be reached in about 17 seconds. Top speed is about 85 mph.
STYLING ✓✓✓✓☐	Typical modern Ford styling, somewhat more conservative than the Detroit versions.
RIDING COMFORT ✓✓✓✓☐	Although both shorter and lighter than the U. S.-built Fords, the Zephyr gives almost as comfortable a ride.
INTERIOR DESIGN ✓✓✓✓☐	Seating arrangements and internal arrangements are excellent. When seating six, it's crowded. But with four passengers, it's roomy. Vision, seating position for drivers and location of controls is very good.
ROADABILITY ✓✓✓✓☐	Roadability is generally very good, although there is some oversteering tendency in sharp cornering.
EASE OF CONTROL ✓✓✓✓☐	Familiar gearshift pattern, positive, smooth, brake action and clutch operation make the Zephyr easy for anyone to drive.
ECONOMY ✓✓✓✓☐	Has exceptional economy for a six-passenger car. Will operate about 23 miles per gallon under average driving conditions. Two or three additional miles per gallon should be made with overdrive equipped car.
SERVICEABILITY ✓✓✓✓✓	Very good. All parts and components easy to reach.
DURABILITY ✓✓✓✓☐	Judged on appearance and performance, the new '56 Zephyr should hold up as well as the earlier model which has an exemplary record of trouble-free operation.
VALUE PER DOLLAR ✓✓✓✓☐	The Zephyr is a well-designed, sturdily-built, useful unit of transportation for those desiring small car attributes neatly balanced with big car capacity and performance.

FORD ZEPHYR

Port of Entry $2100

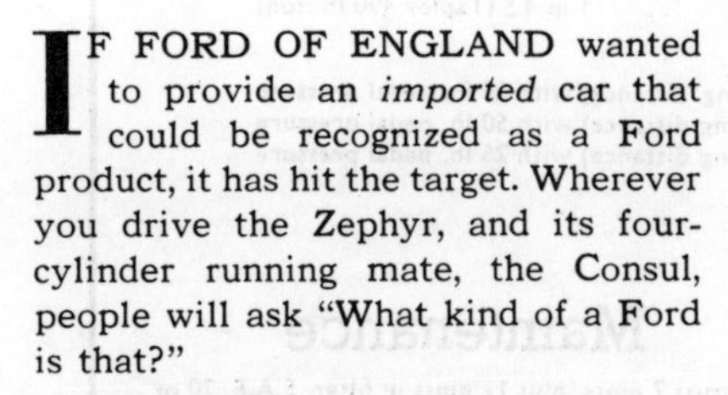

IF FORD OF ENGLAND wanted to provide an *imported* car that could be recognized as a Ford product, it has hit the target. Wherever you drive the Zephyr, and its four-cylinder running mate, the Consul, people will ask "What kind of a Ford is that?"

The family resemblance to the native Michigan product is very strong. With the exception of its grille, taillight assembly and the lack of a wrap-around windshield, the Zephyr could very well pass as a scaled-down '56 domestic Ford.

And the similarity is more than sheet-metal deep too. Almost any local mechanic who has worked on current Ford sixes will be on familiar ground when he pokes under the Zephyr's hood. Only the unique sliding-pillar front suspension with its integral shock absorber-kingpost assembly will seem strange.

The Zephyr has seats almost as wide as the domestic Ford and will accommodate six passengers almost as easily, although the shorter wheelbase (107") does not provide as much legroom for rear seat passengers. Trunk space is a bit smaller.

Performancewise the Zephyr comes closer to the acceleration, cruising speed and hill-climbing ability that U. S. car owners have become accustomed to than most foreign cars.

An important feature of the Zephyr (and all other English-built Fords) is its one-piece body-and-frame construction, which offers a distinct advantage over its U. S. counterpart. The car is rock-solid and virtually rattle free. This type of unit construction makes for a potentially quieter-running car than conventional separate body-and-frame construction. However, to take full advantage of it, the Zephyr (and Consul) need more undercoating and better soundproofing than is provided.

U. S. drivers will be completely at home the moment they slip behind the Zephyr's wheel. It has a three-speed and reverse transmission operated by a smooth-working conventional column lever. Overdrive is available as an optional extra and is well-worthwhile as it increases the flexibility of the engine's performance throughout the 30 to 60 mph speed range. Also it makes for more economy.

Pricewise, the car comes in almost direct competition to the similar model with six-cylinder engine produced in Detroit. Ford Motor Company is counting on the advantages of maneuverability, handling ease and fuel economy to win the Zephyr its share of the lightweight car market in this country.

If the trend toward smaller, more agile automobiles keeps growing, Ford will be ready to cash in with a car that can be built either in Great Britain or Michigan or both. ●

The Motor Road Test No. 9/56 (Continental)

Make: Ford **Type:** Zodiac Mk. II.
Makers: Ford Motor Co. Ltd., Dagenham, Essex.

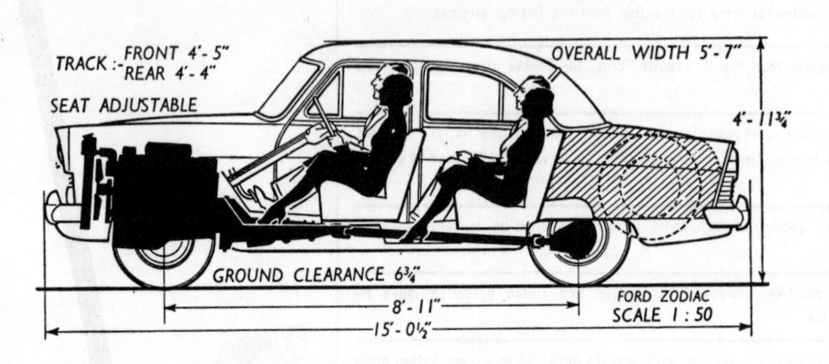

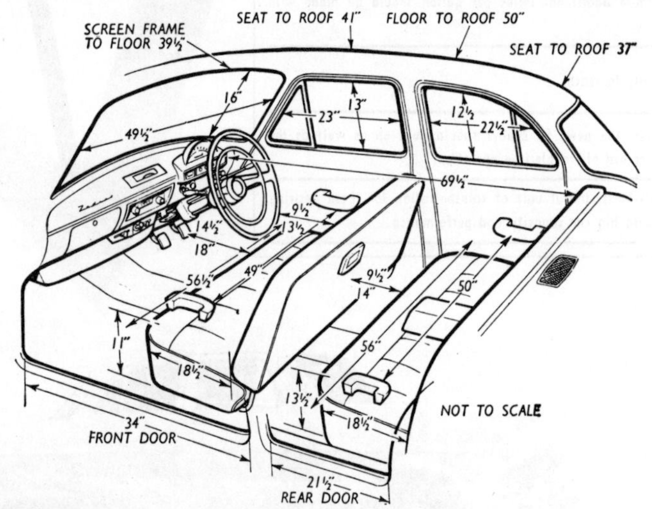

Test Data

CONDITIONS. *Mild, dry weather with little wind (Temperature 40°-50° F., Barometer 29.9-30.0 in. Hg.). Smooth concrete road surface. Premium-grade pump fuel.*

INSTRUMENTS

Speedometer at 30 m.p.h.	5% fast
Speedometer at 60 m.p.h.	5% fast
Distance recorder	5% fast

MAXIMUM SPEEDS

Flying Quarter Mile

Mean of Four Opposite Runs	87.9 m.p.h.
Best Time equals	90.5 m.p.h

Speed in gears

Max. speed in 2nd gear	62 m.p.h.
Max. speed in 1st gear	36 m.p.h.

FUEL CONSUMPTION

31.0 m.p.g. at constant 30 m.p.h.
29.0 m.p.g. at constant 40 m.p.h.
25.5 m.p.g. at constant 50 m.p.h.
22.0 m.p.g. at constant 60 m.p.h.
19.0 m.p.g. at constant 70 m.p.h.
16.0 m.p.g. at constant 80 m.p.h.
Overall consumption for 1,553 miles, 72.3 gallons, = 21.5 m.p.g. (13.1 litres/100 km.)
Fuel tank capacity 11 gallons

ACCELERATION TIMES Through Gears

0-30 m.p.h.	4.6 sec
0-40 m.p.h.	7.4 sec.
0-50 m.p.h.	11.3 sec.
0-60 m.p.h.	17.1 sec.
0-70 m.p.h.	24.3 sec.
0-80 m.p.h.	37.8 sec.
Standing Quarter Mile	20.9 sec.

ACCELERATION TIMES on Two Upper Ratios

	Top	2nd
10-30 m.p.h.	8.1 sec.	4.5 sec.
20-40 m.p.h.	8.0 sec.	5.2 sec.
30-50 m.p.h.	8.4 sec.	6.2 sec.
40-60 m.p.h.	10.1 sec.	10.0 sec.
50-70 m.p.h.	12.9 sec.	—
60-80 m.p.h.	20.7 sec.	—

WEIGHT

Unladen Kerb weight	23¾ cwt.
Front/rear weight distribution	56½/43½
Weight laden as tested	27¼ cwt.

HILL CLIMBING (at steady speeds)

Max. gradient on top gear	1 in 7.3 (Tapley 305 lb./ton)
Max. gradient on 2nd gear	1 in 4.5 (Tapley 490 lb./ton)

BRAKES at 30 m.p.h.

0.95g retardation	(= 31¾ ft. stopping distance) with 85 lb. pedal pressure
0.65g retardation	(= 46½ ft. stopping distance) with 50 lb. pedal pressure
0.30g retardation	(= 100 ft. stopping distance) with 25 lb. pedal pressure

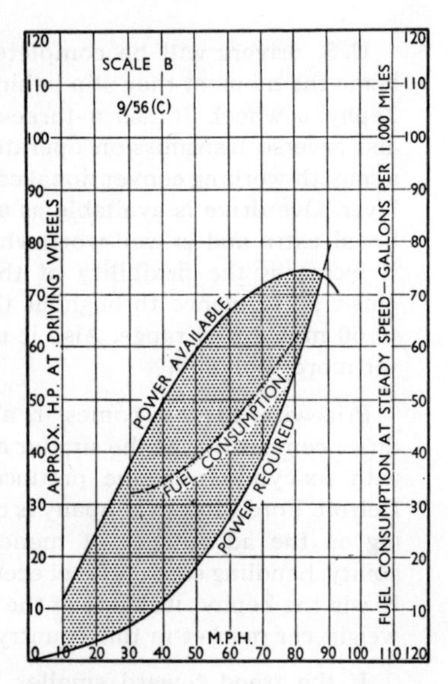

Drag at 10 m.p.h. 54 lb.
Drag at 60 m.p.h. 168 lb.
Specific Fuel Consumption when cruising at 80% of maximum speed (i.e. 70.3 m.p.h.) on level road, based on power delivered to rear wheels .. 0.75 pints/b.h.p./hr.

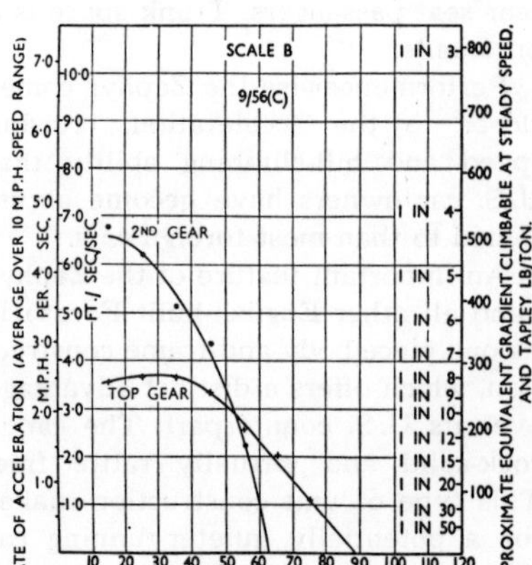

Maintenance

Sump: 7 pints, plus 1½ pints in filter, S.A.E. 20 or 20W for temperate summer or winter. **Gearbox:** 2½ pints, S.A.E. 80 EP gear oil. **Rear Axle:** 2½ pints S.A.E. 90 hypoid oil. **Steering gear:** S.A.E. 80 EP gear oil. **Radiator:** 22½ pints (2 drain taps). **Chassis Lubrication:** Every 1,000 miles, by grease gun to 11 points, by oil gun to 2 points. **Ignition timing:** (Premium fuel) 8° b.t.d.c. **Spark plug gap:** 0.032 in. **Contact breaker gap:** 0.014-0.016 in. **Valve timing:** I.O., 17° b.t.d.c.; I.C., 51° a.b.d.c.; E.O., 49° b.b.d.c.; E.C., 19° a.t.d.c. **Tappet clearances:** (Hot) Inlet and exhaust 0.014 in. **Front wheel toe-in:** 1/16 in. to ⅛ in. **Camber angle:** ½° to 2°. **Castor angle:** +½° to −½°. **King pin inclination:** 4° to 5°. **Tyre pressures:** Front and Rear, 24 lb. **Brake fluid:** EnFo. **Battery:** 12 volt, 57 amp.-hr.

The FORD Zodiac Mk. II

A Fast yet Easy-to-drive New Six-seater from Dagenham

LOW and wide, the latest Zodiac is a car of striking appearance, able to seat 6 people, and with reasonably good traction for motoring off the beaten track.

In Brief

Price: £645 plus purchase tax £323 17s. 0d. equals £968 17s. 0d.

Capacity	2,553 c.c.
Unladen kerb weight ...	23¾ cwt.
Fuel consumption	21.5 m.p.g.
Maximum speed	87.9 m.p.h.
Maximum top gear gradient	1 in 7.3
Acceleration:	
10-30 m.p.h. in top ...	8.1 sec.
0-50 m.p.h. through gears	11.3 sec.

Gearing: 18.3 m.p.h. in top at 1,000 r.p.m.; 35.1 m.p.h. at 1,000 ft. per min. piston speed.

GROWING up is a process which happens to cars as well as to people. Larger and more powerful than its predecessors, without differing greatly from them in general mechanical layout, the latest Ford Zodiac is an altogether more mature car, the merits of which are such that its success seems assured.

With 13% more engine size and 21% more power allowing a 3.90/1 rear-axle ratio to replace the former 4.44/1, with both the wheelbase and track increased by 3 inches, and with a slightly higher price, the Zodiac II has grown up to such an extent that it should rightly be judged on its own merits, and not compared at too great length with the preceding model. To dispose of comparisons straight away, there is 3-6 inches more internal width plus useful extra leg room; top-gear acceleration is 7% faster in the 10-30 m.p.h. range and 12% faster between 40 and 60 m.p.h.; through-the-gears acceleration from a standstill is 25% faster to 50 m.p.h. and 24% faster to 70 m.p.h.; maximum speed is greater by 10%; as the almost inevitable price of all this, there is a fuel consumption increase of only 1-1½ m.p.g. at any steady cruising speed between 30 m.p.h. and 70 m.p.h., and a slight increase in overall fuel consumption.

In view of the American connections of the Ford Motor Company, it is not altogether surprising that the most luxurious model in the Dagenham-built range shows some transatlantic characteristics. After living with the Zodiac for two weeks, however, we felt that it incorporated an exceptionally happy blend of virtues, having the straightforward layout and sensible ground clearance which once was typical of American cars, the much improved road holding and exceptionally good all-round view which are a novelty on the latest models from U.S.A., and the combination of broad bodywork with lively top-gear performance which has always been favoured in America.

Both front and rear seats of the Zodiac have been designed to accommodate three people when required, the hump over the gearbox and propeller shaft not being so large as to be a serious cause of discomfort. The virtual impossibility of reaching an internal near-side door or window handle when sitting in the driving seat emphasizes that this is a much wider car than are most European designs. At the back, side armrests are supplemented by a folding central armrest, a feature which would also be more than welcome in the front of the car, where a passenger sadly lacks lateral support during even moderately brisk cornering, but the side armrest can actually be a nuisance to the driver. Although longer than the previous Zodiac, this car still fails to make a very long-legged driver really comfortable, the range of rearward seat adjustment being limited, and some drivers contacting the steering wheel rim with their right knee when moving a foot on to the brake pedal. Even with the driving seat set fully back, leg

ACCESS to the power unit for routine maintenance work is easy, the brake fluid reservoir, battery, ignition distributor cap, oil filler, carburetter, fuses and screen spray water bottle appearing in this under-bonnet view.

LUGGAGE accommodation is provided on a generous scale, and there is a tool compartment behind the spare wheel. The rear number plate is on a hinged panel which may be pulled open to reveal the petrol filler.

HOODED headlamps and separate red tail lamps and amber direction indicator flashers set high in the rear wings give the Zodiac a long look, to which white-wall tyres and two-colour paintwork draw attention. Four doors give easy entry, and the low bonnet and thin screen pillars provide excellent driving vision.

room in the rear compartment is comfortably adequate for a man of average height, although slight re-shaping of the front-seat frame could advantageously eliminate a hard edge in immediate proximity to a passenger's shins.

From the driving seat the view forwards down a long bonnet is impressive, but as this bonnet droops down to a low nose and is flanked by the sharp ridges of fully visible front wings driving vision is excellent and accurate manoeuvring close to obstacles is easy. Praise is also due for the manner in which sturdy windscreen pillars have been arranged edge-on to the driver's line of vision, so as to be as inconspicuous as possible. The huge rear window and high rear wings make reversing unusually easy. Hanging pedals leave adequate footroom below them, although the dip switch comes almost too naturally where the left foot would otherwise rest.

Switch starting, by turning the ignition key beyond the "on" position, eliminates one control knob—as in spring weather the engine started readily from cold without the choke if the accelerator pedal is "pumped," many owners will use the choke only to secure a fast-idle during the warming-up period. Concentration upon making this an economical car for its size had apparently led to use of lean carburetter settings, which do not always give smooth pulling at low r.p.m. until the engine has been warmed up fully. All the auxiliary controls are on the facia panel, behind the steering wheel, and so only moderately accessible. The main lighting switch is twisted for parking lamps, then pulled out for headlamps, a sensible arrangement, but its location alongside the similar knob which is twisted to switch the screen wipers on or off involves some risk of confusion. Vacuum operated, the screen wipers cover overlapping arcs of useful area, and their speed can be regulated, a vacuum pump ensuring that although they slow down on hills they do not stop. Instruments comprise a speedometer and distance recorder (both of which exaggerated by 5%), an ammeter, a vague fuel contents gauge, and a clock.

Outside the car, the engine is quite audible mechanically and has a marked exhaust note, but to a driver and his passengers neither sound is perceptible and the tick-over is delightfully smooth. First impressions are that the engine only develops its best torque above 30 m.p.h. in top gear, but the figures on the data page show that there is in fact quite exceptionally vigorous top-gear acceleration at any speed between 15 and 75 m.p.h. There are even smoother 6-cylinder engines than this one in respect of low-speed pulling, but the standard of flexibility is commendably high. There is that pleasing virtue of a large car, the ability to pick up a full load of passengers and still handle and perform almost as nimbly as when the driver is alone.

Gearbox Comment

A satisfactorily smooth steering-column control is used for a three-speed gearbox, which has good synchro-mesh on the upper two ratios, first gear also being reasonably easy to engage when the car is moving slowly. The middle gear ratio is quiet and provides outstandingly quick acceleration from very low speeds, being genuinely useful up to over 50 m.p.h. for overtaking other traffic and allowing 60 m.p.h. to be reached on special occasions. A lazy driver can quite happily start in first gear, and at 15-20 m.p.h. make a single gear change directly into top gear. The ability to accelerate from rest to 50 m.p.h. in 11.3 seconds implies superiority over most European cars in respect of get-away from traffic checks. On the test car, slight clutch drag and a throttle which was reluctant to close completely sometimes made it hard to engage first gear silently before starting from rest.

Our test model was not equipped with the semi-automatic overdrive which is an optional extra, but although the engine was audible during acceleration it was very quiet during part-throttle cruising at a genuine 75-80 m.p.h. As so often, opening the hinged panels on the front windows caused wind noise at high speeds and let in rain on wet days, but any of the four main windows could be opened without causing undue draught or noise. No change of rear axle gearing accompanies fitting of an overdrive to this model, and whilst its installation would no doubt improve fuel economy and give even quieter cruising the maximum speed attainable on the level would almost certainly fall below that attained in the direct top gear.

Experience with prototypes in both new and much-used condition suggests that the Zodiac tested was, at about 2,000 miles, perhaps too new to show its best handling qualities. Precise and free from kick-back, the steering had just enough self-centring action to make it sensitive to tramlines or ridged concrete roads, but some friction (which seemed less evident on a car which had run upwards of 50,000 miles) induced a faint trace of wander at low speeds. The lock is very much better than on the previous Zephyr and Zodiac models. In general, this car sweeps through main-road curves at its naturally fast cruising gait in delightful fashion, setting up high average speeds quite easily, but on sharp corners where it is more likely to be handled vigorously it is a little less tidy—although with more of the weight on the driven rear wheels, it does not easily spin its tyres as did the preceding Zodiac. At around 70 m.p.h. some shake of the steering column was evident on the test model.

Substantially conventional suspension is used, quite firm springs and dampers properly matched together giving a generally good ride over all kinds of surfaces, for rear-seat passengers as well as for the

BENCH type seats have a central armrest at the rear but not, unfortunately, at the front. A map shelf and glove locker face the passenger.

driver, bumps being felt but never exaggerated. The amount of body roll during cornering is quite modest, thanks partly to the wide track, and there is no great amount of road noise transmitted into the body, whilst it needs quite vigorous cornering to make the tubeless tyres squeal in protest. The latest braking system, with very wide front drums of composite construction, works well with very light pedal pressures in normal driving and is free from overheating and fade even when the car is driven at well above average speeds. The pull-out handbrake is mounted at a convenient angle, and so works more easily than many of its kind.

Gay two-colour exterior paintwork and interior decoration are offered on the Zodiac, with the option of black paintwork and a brown interior. Bright colour schemes do not appeal to all British motorists (the Zephyr is mechanically similar but is rather more plain in style), but internally and externally the decorative treatment of this latest Zodiac seems much better designed than was that of its predecessor. Sensibly, the roof has a washable plastics lining, but the imitation leather above the instrument panel showed signs of staining easily. Carrying capacity for odds and ends is fairly good, with a small glove locker the key of which is foolishly the same as the ignition key, a map shelf of modest depth facing the passenger, and another shelf behind the rear seat.

Generous capacity for luggage is provided, in a rear locker which, although peculiarly shaped (it extends out into the rear wings behind the wheels), has large sections of flat floor—the luggage locker catch which requires firm pressure on a shrouded button to release it might prove hard upon long feminine finger-nails! Beneath the luggage locker is an 11-gallon fuel tank, the central filler of which will take petrol at full speed from an electric pump, but a tethered fuel cap which cannot be lost as well as some catch to hold back the spring-loaded panel giving access to the fuel filler would be welcome. Both the luggage locker lid and the bonnet panel are spring counter-balanced to open easily and remain open, and beneath the bonnet the oil dipstick, o.h.v. rocker cover, ignition contact breaker, fuel filter, and carburetter are readily accessible.

Interior heating is standardized on the Zodiac, and proved both effective and easy to regulate: a detail fault which should be simple to cure was incomplete mixing of adjacent streams of heated and unheated air entering the car when partial heating only was needed. A new Enfo radio of Ekco manufacture is an optional extra, a set which has no press-buttons or other pre-set tuning, but proved quite easy to tune manually on two wavebands and disclosed very good power and tone in relation to its modest cost. Fog lamps, exterior mirrors and reversing lamps are not now standardized fittings.

Fundamentally, the appeal of the latest six-cylinder Fords is based on value for money; on the provision of six-seat dimensions and vivid acceleration up to nearly 90 m.p.h. at a price, even including Britain's purchase tax, of under £1,000. Working on the lines of preceding models, which had been very popular and successful, the Ford Motor Company have been able to obtain a very high standard in respect of the intangibles which make a car pleasant, as well as offering the exceptional value which has for so many years been associated with the name of the late Henry Ford.

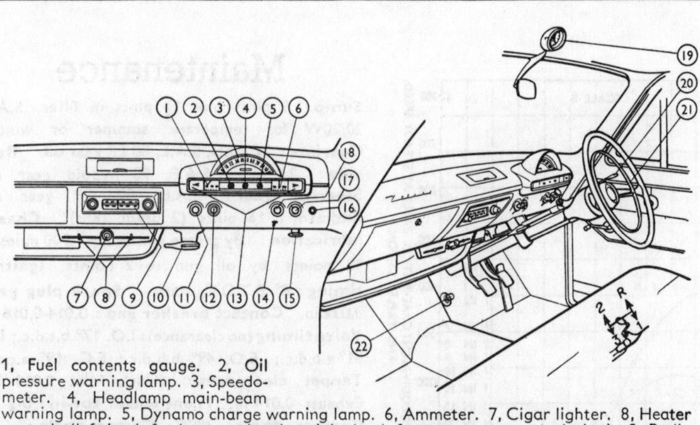

1, Fuel contents gauge. 2, Oil pressure warning lamp. 3, Speedometer. 4, Headlamp main-beam warning lamp. 5, Dynamo charge warning lamp. 6, Ammeter. 7, Cigar lighter. 8, Heater controls (left knob for heat or de-mist, right knob for temperature regulation). 9, Radio (optional extra). 10, Hand brake (pull). 11, Ignition key (turn left for accessories only, right for ignition on, farther right against spring to operate starter). 12, Choke (pull). 13, Direction indicator warning lamps. 14, Instrument lighting switch. 15, Bonnet release (pull). 16, Windscreen washing spray (press). 17, Windscreen wiper control (turn left). 18, Lighting switch (turn right for sidelamps, then pull out for headlamps). 19, Clock. 20, Direction indicators (turn with wheel). 21, Horn ring (press). 22, Headlamp dip switch (press).

Mechanical Specification

Engine

Cylinders	6
Bore	82.55 mm.
Stroke	79.5 mm.
Cubic capacity	2,553 c.c.
Piston area	49.74 sq. in.
Valves	Pushrod o.h.v.
Compression ratio	7.8/1
Max. power	86 b.h.p.
at	4,200 r.p.m.
Piston speed at max. b.h.p.	2,190 ft. per min.
Carburetter	Zenith 36WIA downdraught
Ignition	12-volt coil
Sparking plugs ...	14 mm. Champion N8B
Fuel pump	AC mechanical, with vacuum pump
Oil filter	AC full-flow

Transmission

Clutch	Single dry plate
Top gear (s/m)	3.90
2nd gear (s/m)	6.40
1st gear	11.08
Propeller shaft	Open
Final drive...	Hypoid bevel
Top gear m.p.h. at 1,000 r.p.m.	18.3
Top gear m.p.h. at 1,000 ft./min. piston speed	35.1

Chassis

Brakes ...	Girling hydraulic, 2 l.s. front
Brake drum diameter	9 in.
Friction lining area	147 sq. in.
Suspension:	
Front	Telescopic coil-spring i.f.s., with anti-roll torsion bar
Rear	Semi-elliptic
Shock absorbers:	
Front	Incorporated in i.f.s.
Rear	Lever-arm hydraulic
Tyres	6.40—13

Steering

Steering gear	Worm and peg
Turning circle (between kerbs):	
Left	34¾ feet
Right	33¾ feet
Turns of steering wheel, lock to lock	3⅓

Performance factors (at laden weight as tested):

Piston area, sq. in. per ton ...	35.8
Brake lining area, sq. in. per ton	106
Specific displacement, litres per ton mile	3,020

Fully described in *The Motor*, April 11, 1956.

Coachwork and Equipment

Bumper height with car unladen:	
Front (max.) 23 in. (min.) 14 in.	
Rear (max.) 25 in. (min.) 16 in.	
Starting handle	No
Battery mounting ...	On right of engine
Jack	Bevel-geared bipod type
Jacking points ...	One on each side of car, external
Standard tool kit:	Jack, wheelbrace, grease gun, pliers, 2 screwdrivers, 2 double-ended open spanners, 2 double-ended box spanners, spark plug, box spanner, combined tommy bar and drain plug key.
Exterior lights:	2 headlamps, 2 side/direction indicator lamps, 2 stop/tail lamps (rear direction indicators are separate), 2 rear number plate lamps.
Direction indicators ...	Flashing type (white front, amber rear) self cancelling
Windscreen wipers:	Vacuum operated (with engine driven pump), two-blade, self-parking, with screen spray.
Sun vizors	2, universally pivoted
Instruments:	Speedometer with decimal non-trip distance recorder, fuel contents gauge, ammeter..
Warning lights:	Oil pressure, dynamo charge, headlamp main beam, direction indicators.
Locks:	
With ignition key	Ignition, either front door, glove box, luggage locker
With other keys	None
Glove lockers ...	One on facia, with lock
Map pockets ...	Below facia panel, facing passenger
Parcel shelves	Behind rear seat
Ashtrays	One on facia, one behind front seat
Cigar lighters ...	One on facia panel
Interior lights ...	One on centre pillar, with courtesy switches
Interior heater	Fresh air type heater and de-mister (air intake below windscreen)
Car radio	Optional extra EnFo
Extras available:	Overdrive (also EnFo approved accessories for fitting by dealers).
Upholstery material ...	Leather facings with matching plastics elsewhere
Floor covering	Pile carpet
Exterior colours:	Black or five two-tone colour schemes.
Alternative body styles ...	Zephyr saloon

The Motor Road Test No. 12/56

Make: Ford. **Type:** Consul Mk. II Saloon

Makers: Ford Motor Co. Ltd., Dagenham, Essex.

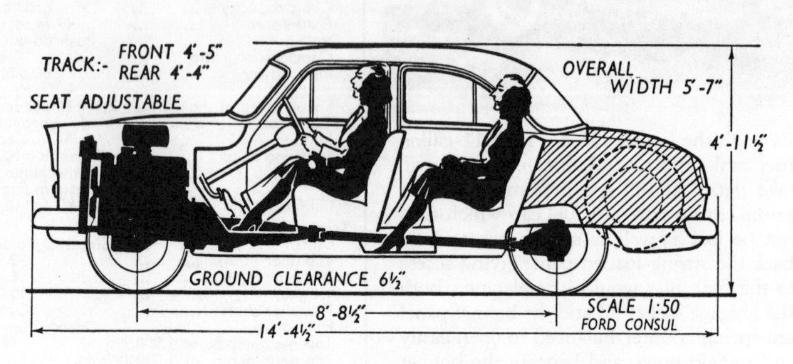

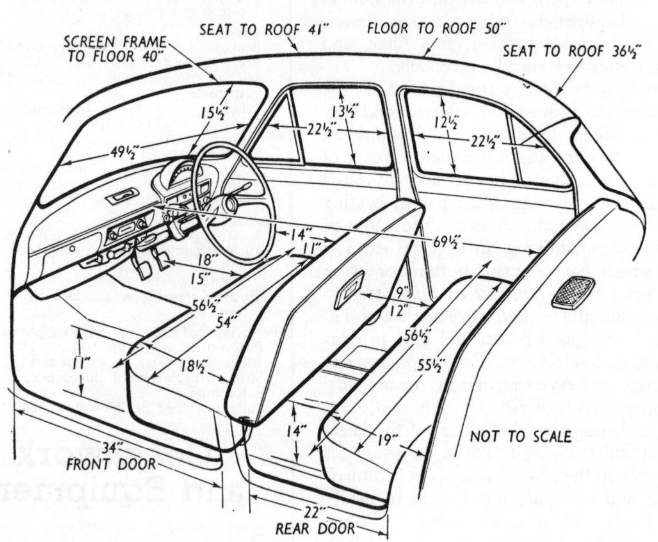

Test Data

CONDITIONS. *Mild, dry weather with little wind (Temperature 58°-63°F., barometer 29.9-30.0 in. hg.). Smooth tarred road surface. Premium-grade pump fuel.*

INSTRUMENTS

Speedometer at 30 m.p.h.	2% fast
Speedometer at 60 m.p.h.	2% fast
Distance recorder	2% fast

MAXIMUM SPEEDS

Flying Quarter Mile

Mean of four opposite runs	79.3 m.p.h
Best time equals	82.6 m.p.h

"Maximile" Speed (Timed quarter mile after one mile accelerating from rest)

Mean of four opposite runs	77.6 m.p.h
Best time equals	79.6 m.p.h

Speed in gears

Max. speed in 2nd gear	59 m.p.h.
Max. speed in 1st gear	34 m.p.h.

FUEL CONSUMPTION

37.0 m.p.g. at constant 30 m.p.h.
34.5 m.p.g. at constant 40 m.p.h.
31.0 m.p.g. at constant 50 m.p.h.
27.0 m.p.g. at constant 60 m.p.h.
20.5 m.p.g. at constant 70 m.p.h.

Overall consumption for 1143.5 miles 51.75 gallons = 22.1 m.p.g. (12.8 litres/100 km).
Fuel tank capacity 11 gallons.

ACCELERATION TIMES Through Gears

0-30 m.p.h.	5.6 sec.
0-40 m.p.h.	9.9 sec.
0-50 m.p.h.	14.8 sec.
0-60 m.p.h.	23.2 sec.
0-70 m.p.h.	36.9 sec.
Standing Quarter Mile	23.0 sec.

ACCELERATION TIMES on Two Upper Ratios

	Top	2nd
10-30 m.p.h.	11.1 sec.	5.9 sec.
20-40 m.p.h.	11.0 sec.	6.7 sec.
30-50 m.p.h.	11.8 sec.	8.3 sec.
40-60 m.p.h.	15.1 sec.	—
50-70 m.p.h.	22.7 sec.	—

WEIGHT

Unladen kerb weight	22 cwt.
Front/rear weight distribution ..	55/45
Weight laden as tested	25½ cwt.

SCALE B 12/56

APPROX. H.P. AT DRIVING WHEELS

FUEL CONSUMPTION AT STEADY SPEED—GALLONS PER 1,000 MILES

POWER AVAILABLE

FUEL CONSUMPTION

POWER REQUIRED

M.P.H.

Drag at 10 m.p.h 49 lb
Drag at 60 m.p.h. 157 lb.
Specific Fuel Consumption when cruising at 80% of maximum speed (i.e. 63.4 m.p.h.) on level road, based on power delivered to rear wheels .. .. .. 0.71 pints/b.h.p./hr

HILL CLIMBING (At steady speeds)

Max. gradient on top gear	1 in 10.6 (Tapley 210 lb./ton)
Max. gradient on 2nd gear	1 in 6 (Tapley 370 lb./ton)

BRAKES at 30 m.p.h.

0.87g retardation ..	(=34½ ft. stopping distance) with 75 lb. pedal pressure
0.70g retardation ..	(= 43 ft. stopping distance) with 50 lb. pedal pressure
0.27g retardation ..	(=111 ft. stopping distance) with 25 lb. pedal pressure

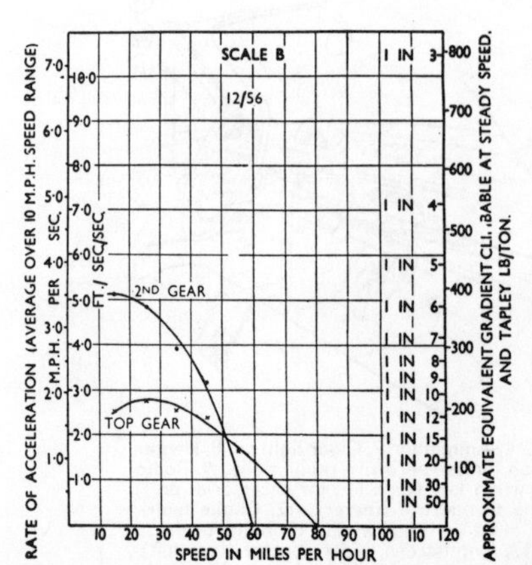

Maintenance

Sump: 6 pints plus 1½ pints in filter, S.A.E. 20/20W for temperate summer or winter. **Gearbox:** 2½ pints, S.A.E. 80 EP gear oil. **Rear axle:** 2½ pints, S.A.E. 90 hypoid gear oil. **Steering gear:** S.A.E. 80 EP gear oil. **Radiator:** 16 pints (2 drain taps). **Chassis lubrication:** By grease gun every 1,000 miles to 12 points, by oil gun to 2 points. **Ignition timing:** 8° B.T.D.C. static. **Spark plug gap:** 0.032 in. **Contact breaker gap:** 0.014-0.016 in. **Valve timing** (no clearance): I.O. 17° b.t.d.c.; I.C. 51° a.b.d.c.; E.O. 49° b.b.d.c.; E.C. 19° a.t.d.c. **Tappet clearances:** (Hot), Inlet 0.014 in., Exhaust 0.014 in. **Front wheel toe-in:** 0-⅛ in. **Camber angle:** ½°-2°. **Castor angle:** +½°-−½°. King pin inclination 4°-5°. **Tyre pressures:** Front and rear, 28 lb. **Brake fluid:** EnFo ME3833-E. **Battery:** 12 volt, 45 amp.-hr.

The FORD Consul Mk. II Saloon

A Well-styled and Spacious Car of Low Price and High Performance

STYLISH in the best sense, the new Consul presents a valuable blend of attractive appearance and practical layout. Occupants are offered notably good all-round visibility.

SMART appearance and ample carrying capacity are the most obviously attractive features of the new Ford Consul.

The general outline of the car shows the swing in contemporary design from the curvilinear to the rectilinear, for just as the day of "razor-edge" styling is dead so is the day of the bulbous-looking car coming to an end. Certainly the new Consuls have an agreeably lean and sharp look, which is emphasized by the peaks of the headlamp hoods at the front and the suggestion of tail fins at the back.

If one opens the doors the evidence of thought in the styling studio is even stronger. The seats are of lightly coloured woven plastics material buttoned to reduce, although it must be admitted, not wholly to prevent, wrinkling in use; the facia is also plastics-covered and has a smart assembly for the instruments which includes a deep hood for the speedometer; the interior of the roof is easily washable, the door handles and window winders have an agreeable appearance and the simple yet elegant steering wheel has an unobtrusive coloured centre. In other words, although the Consul is a car of comparatively low price, it is far from being austere or drab. On the contrary it is gaily coloured and well proportioned.

It is also exceptionally roomy. Something of this can be seen from the photographs and any impression they give is reinforced by a study of the dimensional body drawings which show that there are 54 inches across the front and 55½ inches across the rear seats. The carriage of six persons is therefore not an uncomfortable possibility, acceptable only as an emergency over short distances, but a quite reasonable proposition especially as wheel arches, propeller shaft tunnel, and clutch housing make but small inroads upon the platform area. Similarly, although the spare wheel is mounted naked and unashamed in the 20-cu. ft. rear luggage locker, the space remaining for suitcases is far above that normal to this class of car.

The fact that the luggage locker lid is counter-balanced draws attention to the many practical features to be found in the Consul which certainly cannot be criticized on the ground that styling conflicts either with engineering or commonsense. A valuable minor item is an ability to unlock both front doors, either of which can be previously sealed from the inside by the driver; and a major attraction is the visibility conferred by the full width rear window, the equally wide windscreen, and the combination of low scuttle with falling bonnet line.

The indentation in the bonnet top to match the scuttle air intake is attractive, the prominence of the front wings helps to position the car on the road and the large rear-view mirror gives the driver a commanding view of what is behind him.

Carriage of personal effects is well taken care of by a cubbyhole on the facia, of which the door has to be locked to maintain it in position, and by an open parcel shelf placed immediately below it. Relatively large objects may be placed upon a broad shelf between the rear seat and the bottom edge of the sweeping rear window.

INTERIOR design is harmonious and sensible, instruments being easy to read and hand controls easy to reach, but pendant clutch and brake pedals do not mate well with a low, organ-type throttle pedal, and the driving position could be made more comfortable.

In Brief

Price: £520 plus purchase tax £261 7s. 0d. equals £781 7s. 0d.

Capacity	1,702 c.c.
Unladen kerb weight ...	22 cwt.
Fuel consumption	22.1 m.p.g.
Maximum speed	79.3 m.p.h.
"Maximile" speed	77.6 m.p.h.
Maximum top gear gradient	1 in 10.6
Acceleration:	
10-30 m.p.h. in top ...	11.1 sec.
0-50 m.p.h. through gears	14.8 sec.

Gearing: 16.6 m.p.h. in top at 1,000 r.p.m.; 31.8 m.p.h. at 1,000 ft. per min. piston speed.

REVERSING and parking are simple when such a large rear window is allied to prominent wings. The petrol filler lies behind the spring-loaded number-plate.

The interior ventilation is highly efficient and easily controlled. A lever moving in a horizontal plane beneath the facia panel to the left of the centre-line of the car determines the distribution of the incoming air, from all to the body to the opposite extreme of all to the slots at the base of the windscreen. A similar lever placed right of the centre-line of the car controls air temperature from cold to hot with the option of bringing in the booster fan at either end of the scale to supplement the ram effect which normally suffices when travelling at over 30 m.p.h. There are large and rigidly mounted ashtrays, the switches are clearly distinguishable and quickly accessible, and although the instruments may easily be read at night they do not promote dazzle or reflection.

Less praiseworthy is the hinged and sprung rear number plate. This conceals the petrol filler but is apt to graze the skin of the petrol-pump attendant; the steering wheel is too close to the seat even when the latter is put right back, and the position of the pedals brings them rather high from the floor and prevents immediate transfer of the right foot from the accelerator to the brake. The driving position as an entity was in fact commented upon unfavourably by most people who drove the car, and tall men might experience real difficulty in comfortably accommodating themselves.

Praiseworthy Performance

We need not speak in great detail about the performance of the car since the figures speak for themselves. They present a closer analogue to the earlier Zephyr than to the preceding Consul; in fact the maximum speed and 0-50 m.p.h. acceleration figures are almost exactly the same as those recorded on the earlier 6-cylinder types.

Compared to the Mark I Consul, the top-gear acceleration and hill climbing show no major improvement, but the latest car is, of course, considerably larger and roomier (although interestingly enough only 25 lb. heavier), and is also appreciably higher geared, engine speed at 60 m.p.h. having been reduced from 4,000 r.p.m. to 3,600 r.p.m.

Although the Consul will reach maximum, and sustain cruising speeds, equivalent to those put up by many 6-cylinder models of greater engine size, it is idle to expect any 4-cylinder engine above 1½-litres swept volume to compare with a six in quietness and smoothness. From the vibration point of view the best that the designer can do for the four is to choose the point in the speed range where its obtrusiveness will be least unwelcome, and on the Ford this is between 15 and 25 m.p.h. The power unit becomes progressively quieter and sweeter up to 50-55 m.p.h., after which smoothness is maintained, but there is naturally a steadily increasing awareness of the engine from the noise point of view. Indeed, the car cannot lay claim to high standards of quietness although it may well be that some of the squeaks and rattles present on the test model (notably one caused by insufficient clearance between the exhaust pipe and the petrol tank) will be eliminated with greater production experience.

These minor defects were of little account compared to the high performance of the car which sweeps along at between 60 and 70 m.p.h. in such a manner that excellent average speeds can be put up. Matching this high cruising speed are the brakes. These show an average relation of stopping power to pedal pressure but have an unusually large figure of sq. in. per laden ton which gives them stability when used hard and frequently, and an ability to be run for well over 2,000 miles without adjustment even with the driving normal to *The Motor* road test. They are free of the modern vices of squeak or judder and, with an exception that will be mentioned later, can be given full marks.

The suspension system also merits high praise under two conditions frequently experienced. On smooth main roads cornering power is good and the roll angle by no means excessive. Hence there need be but little diminution in cruising speed on sinuous sections, although passengers may complain that the side forces to which they are exposed are difficult to resist owing to a complete absence of armrests or hand grips. In the opposite extreme of really rough roads the Consul springing system has exceptional merit, for really large bumps are swallowed up as if by magic and with big wheel motions the damping is highly effective. But on fast wavy surfaces there is a good deal of undamped vertical

AMPLY WIDE, the bench seats can each carry three people, and the front-hinged doors swing open to a usefully wide angle. Trim and fittings are well designed.

- - - - - - - Contd.

motion and although this does not degenerate into pitch it can have an adverse effect upon the mind and digestion of the passengers.

Contrasts in Control

The steering, as distinct from the cornering, also has varying degrees of merit. The turning circle is satisfactorily small, the castor return action from large angles of lock is good without excessive turning effort, and all road irregularities are damped out. The car is stable at speed on the straight, partly by reason of an internal damping in the mechanism when at the neutral point. This can be irritating at low speeds and in traffic as it makes it difficult to steer the car within an inch or two sideways and thus causes the driver to hesitate to pass between two objects when the clearance is only three or four inches above the full width of the Ford.

When the roads are wet the normally welcome pronounced under-steering of the car is less desirable, and one may also experience unexpected loss of adhesion by the rear wheels. So if good speeds are to be maintained over slippery road surfaces the driver must be alert and able to exercise a degree of skill. Similarly, on wet roads the front wheels lock first in an emergency and although the car always goes straight on, which is sometimes desirable, steering power is lost. Lastly, comment may be aroused by the fuel consumption figure, which is about the same as that achieved by the earlier model of the Zephyr. It is therefore pertinent to point out that there is little difference between the engine output of the old Zephyr and the new Consul; the maximum speed and all-round performance of the cars are almost identical; and the Consul has a wider, roomier body with the same wind resistance. In these contexts the swept volume of the engine is largely irrelevant, for fuel consumption is determined by the work done during the day. It follows that those drivers who deny themselves enjoyment of the outstanding performance offered by the Consul can claim a reward in improved fuel consumption.

Summing up, the new Ford Consul offers outstanding value for money. For substantially under £800 (with British purchase tax included) it offers the buyer a combination of highly attractive appearance, exceptional spaciousness, excellent visibility, comfort, performance and practical features which places it in the front rank of British volume production cars.

REALLY large quantities of luggage can be swallowed by the boot even though the spare wheel is accommodated therein. The lid is counterbalanced and swings well out of the way when opened.

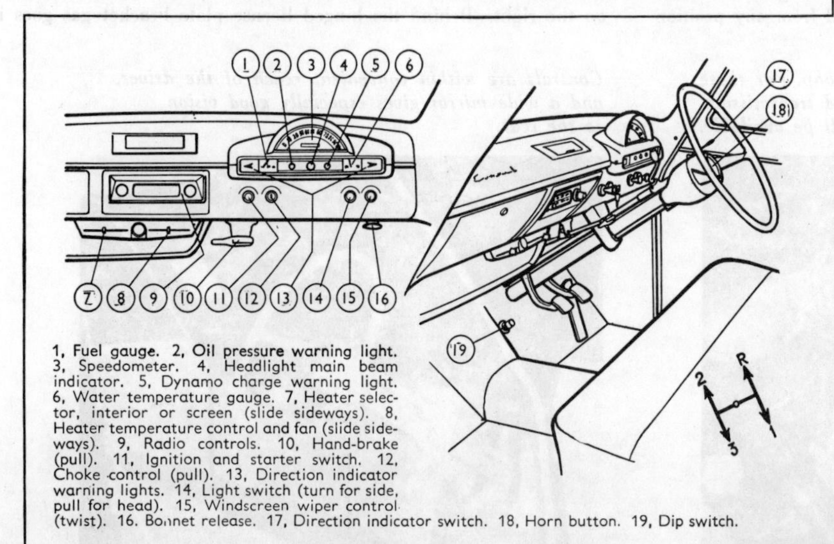

1, Fuel gauge. 2, Oil pressure warning light. 3, Speedometer. 4, Headlight main beam indicator. 5, Dynamo charge warning light. 6, Water temperature gauge. 7, Heater selector, interior or screen (slide sideways). 8, Heater temperature control and fan (slide sideways). 9, Radio controls. 10, Hand-brake (pull). 11, Ignition and starter switch. 12, Choke control (pull). 13, Direction indicator warning lights. 14, Light switch (turn for side, pull for head). 15, Windscreen wiper control (twist). 16. Bonnet release. 17, Direction indicator switch. 18, Horn button. 19, Dip switch.

Mechanical Specification

Engine

Cylinders	4
Bore	82.55 mm.
Stroke	79.5 mm.
Cubic capacity	1,702 c.c.
Piston area	33.16 sq. in.
Valves	Push-rod o.h.v.
Compression ratio	7.8/1
Max. power	59 b.h.p.
at	4,200 r.p.m.
Piston speed at max b.h.p.	2,190 ft. per min.
Carburetter	Zenith 34WIA downdraught
Ignition	12-volt coil
Sparking plugs ...	14 mm. Champion N8B
Fuel pump	Mechanical (with vacuum pump)
Oil filter	A.C. full-flow

Transmission

Clutch ...	Single dry plate, 8 in. dia.
Top gear (s/m)	4.11
2nd gear (s/m)	6.75
1st gear	11.67
Propeller shaft ...	Hardy Spicer open
Final drive	Hypoid bevel
Top gear m.p.h. at 1,000 r.p.m.	16.6
Top gear m.p.h. at 1,000 ft./min. piston speed	31.8

Chassis

Brakes ...	Girling hydraulic, 2 l.s. front
Brake drum diameter	9 in.
Friction lining area	147 sq. in.
Suspension:	
Front	Coil spring i.f.s. with anti-roll torsion bar
Rear	Semi-elliptic
Shock absorbers:	
Front	Incorporated in i.f.s.
Rear	Lever-arm hydraulic
Tyres	5.90—13

Steering

Steering gear	Worm and peg
Turning circle (between kerbs):	
Left	32 feet
Right	31¾ feet
Turns of steering wheel, lock to lock	3⅜

Performance factors (at laden weight as tested):

Piston area, sq. in. per ton	...26
Brake lining area, sq. in. per ton	115
Specific displacement, litres per ton mile	2,415

Fully described in *The Motor*, April 11, 1956.

Coachwork and Equipment

Bumper height with car unladen:
Front (max.) 19 in., (min.) 15½ in.
Rear (max.) 20 in., (min.) 16½ in.

Starting handle	No
Battery mounting	Alongside engine
Jack	Bevel geared bipod type
Jacking points	External, one on each side of body

Standard tool kit: Jack, wheelbrace, tyre lever, pliers, adjustable spanner, sparking plug spanner, 2 d.e. box spanners, combined tommy bar and drain plug key, 2 d.e. open spanners, 2 screwdrivers, Phillips screwdriver.

Exterior lights: 2 headlamps, 2 sidelamps/flashers, 2 stop/tail lamps, number plate lamp.

Direction indicators ...	Flashing type, self cancelling (amber at rear)
Windscreen wipers ...	Vacuum operated 2-blade, self-parking
Sun vizors	Two, universally pivoted

Instruments: Speedometer with non-trip decimal distance recorder, fuel contents gauge, ammeter.

Warning lights: Dynamo charge, oil pressure, headlamp main beam, direction indicators.

Locks:	
With ignition key ...	Ignition, both front doors, glove box, luggage locker
With other keys	None
Glove lockers ...	One on facia, with lock
Map pockets ...	Shelf below facia panel
Parcel shelves	Behind rear seat
Ashtrays	One on facia panel, one behind front seat
Cigar lighters	Optional extra, on facia panel
Interior lights ...	One on centre door pillar, with courtesy switches
Interior heater ...	Optional extra, fresh-air type with screen de-misters
Car radio	Optional extra

Extras available: Radio, heater, screen washing sprays, electric clock, cigar lighter, Borg-Warner overdrive.

Upholstery material	Woven plastic fabric
Floor covering	Rubber
Exterior colours standardized ...	Eight
Alternative body styles ...	2-door, 4/6-seat convertible

Combination of high roll center front suspension and anti-roll bar gives comparatively little tilt on corners.

Windshield and rear window are broadly curved and provide excellent all-around visibility for driver and passengers.

Road Test: FORD CONSUL II

THERE are six members of the English Ford family (Popular, Anglia, Prefect, Consul, Zephyr, and Zodiac), and none bears any resemblance to its American cousins—fortunately. While Dearborn actually boasts of a product "over 17 feet long," Dagenham sensibly offers (in the Consul) an overall length (14 ft. 4.5 in.) that can be conveniently contained in the average home garage and street parking space. There is a look about the new grille on the Consul II that remotely suggests a Thunderbird, but similarity ends there. Far from being "longer, lower, wider," the Consul's moderate dimensions are, nonetheless, more than adequate to handle the transportation needs of a reasonably prolific family on all occasions short of moving-van duty. Although its overall height is only 2.1 in. more than a U.S. Ford (59.3 vs. 57.2), the seat-to-roof space is 41 in. in front and 36.5 in back, and the transmission/driveshaft tunnel, which has turned many a 1957 U.S. hardtop into virtually a four-seater as far as leg room is concerned, obtrudes negligibly in the Consul. Six people really can ride in the car, and a six-footer-plus can drive with fedora on head—an almost forgotten luxury in this country. True, the car will not rumble off from 0 to 60 in ten seconds, but neither will it consume gas at the rate of 15 mpg.

Our Consul test car was supplied by Bob Knapp Motors of Pasadena and was finished in black with light plastic upholstery. Viewed from the outside, the car is pleasantly free of superfluous ornament, has well balanced contours, and is shaped in a body style popular in Detroit half a dozen years ago. The Consult shares its unit-frame body shell with the more expensive Zephyr and Zodiac, but has fewer frills and a four (instead of six) cylinder engine. All four doors open wide for easy entrance and exit, and from any position inside the car there is excellent all-around visibility. A wide, curved rear window and high rear fenders which define the corners of the body combine to facilitate backing and parking; the driver, in fact, has a reassuring feeling at the wheel of the Consul that he can gauge the physical limits of his car with a good deal of precision. The interior is fairly conventional as to layout and materials, and front and rear seats are comfortably wide at 54 and 55.5 inches, respectively. In our test car there were a good many loose ends and "Irish pennants" visible throughout the interior, as if the upholsterers had finished up in a hurry, and there were also more rattles than might be expected in a new car, but judging from other models inspected, neither criticism seems to be characteristic.

Driver controls are conveniently located, with pendant brake and clutch pedals and a mildly "dished" steering wheel that is situated a little too close to the legs for easy exit-and-entrance movement. As one of several good safety features, the emergency brake is mounted under the dash at the right of the steering column where it could be reached if necessary by a passenger. The windshield is gently curved across the middle, more sharply curved near the ends, and in the latter portions there is the inevitable distortion which makes trees curtsey and fences dip when traveling along the highway; the flaw is minor compared to most U. S. "wraparounds" however. The hood slopes down between the front fenders permitting a good view of the road directly in front of the car, and absence of any hood ornament gives a smooth, "customized" look to the front end. At the rear a counterbalanced lid swings up to reveal 20 cu. ft. of trunk space with the spare mounted vertically on the right. Behind the hinged license plate bracket gas goes in.

The overhead-valve four develops 59 bhp. For those not content with a 3-speed transmission, overdrive will be available.

Controls are within convenient reach of the driver, and a wide mirror gives especially good vision to the rear.

New grille styling gives clean, "customized" look.

What Dearborn doesn't have, Dagenham does

Under the hood is a sturdy ohv four displacing 1702cc (103.9 cu. in.) and developing 59 bhp at 4200 rpm. Power is transmitted via an 8 in., single-dry-plate clutch and 3-speed gearbox with the two top gears synchronized. Although Ford has certainly not conceived of the Consul as a "performer," it does well enough under the stopwatch to give a feeling of security in American traffic. Our standing quarter and 0-to-60 mph times were within the same half-second, which means that a Consul owner isn't going to beat any domestic product at the stoplight, but on the other hand he won't be trampled in the rush. Our top speed of just under 80 mph was a long time in coming, but maintaining 70 for highway cruising is no problem. It should be mentioned, however, that, characteristic of its breed, the 4-cyl. engine is not notable for smoothness or quietness under pressure. It actually has a labored sound when pushed hard, even though a piston speed of 2500 ft./min. is just barely obtainable in top gear on a level road.

Perhaps the most winning quality of the Ford Consul for anyone who cares how a car handles is its high degree of controlability and obedience over a wide range of road surfaces. There is some understeer on corners, but body roll is moderate and the suspension (coils with anti-roll bar in front, semi-elliptics in rear) firm enough to avoid wallow at higher speeds. With a light load, some may feel the suspension is a little too firm, since at speeds under 35-40 mph minor shocks (but not major) are transmitted to occupants, but for a 4-door family sedan the ride compromise seems to have far more on the credit than the debit side. The same can be said for the steering: 3.2 turns lock-to-lock combined with a turning diameter of 32 ft. (8 ft. shorter than Dearborn's Ford) gives just the right ease of maneuverability for a 2500 lb. car. Special mention should be given the brakes which. with a total lining area of 147 sq. in., required less than average pedal pressure and showed good resistance to fade. Like most British brakes, however, they were by no means quiet at low speed, especially when inching along in traffic.

We parted from the Consul with much the same feeling we expressed last month about the Simca Versailles: Both are extremely practical cars for moderate-income U, S. families, and both can readily serve either as a full-status family sedan or as a light, economical second car in more affluent households. We also believe that of the British Ford line, the Consul represents the most for the money in matching economy of cost and maintenance against space and performance. And as an added and by no means insignificant-attraction, British Fords are serviceable at selected Ford dealers in this country and use the new international size nuts and bolts, thus rendering innocuous that all-to-familiar bogey that has plagued so many imported car owners who live or travel away from the coastal metropolises. ●

R & T ROAD TEST NO. 124

FORD CONSUL SEDAN

SPECIFICATIONS

List price	$1968
Wheelbase, in.	104.5
Tread, f/r	53/52
Tire size,	5.90-13
Curb weight, lbs.	2540
Distribution, %	53/47
Test weight	2880
Engine	4 cyl, ohv
Bore & stroke	3.25 x 3.125
Displacement, cu in.	103.9
cu cm.	1702.6
Compression ratio	7.80
Horsepower	59
peaking speed	4200
equivalent mph	70
Torque, ft-lbs.	92
peaking speed	2300
equivalent mph	38.2
Gear ratios, overall	
3rd	4.11
2nd	6.75
1st	11.7

CALCULATED DATA

Lbs/hp (test wt.)	48.8
Cu ft/ton mile	75.2
Engine revs/mile	3610
Piston travel, ft./mi.	1880
Mph @ 2500 fpm	79.7

PERFORMANCE, Mph

Top speed, avg.	78.6
best run	79.8
2nd (5800)	59
1st (5800)	34
shift points used	
2nd (5500)	56
1st (5500)	32
Mileage range	23/28 mpg

ACCELERATION, Secs.

0-30 mph	5.5
0-40 mph	9.6
0-50 mph	14.1
0-60 mph	22.5
0-70 mph	34.4
standing start 1/4 mile	22.8

TAPLEY DATA, Lbs/ton

3rd	185 @ 44 mph
2nd	300 @ 31 mph
1st	450 @ 22 mph
Total drag at 60 mph, 148 lbs.	

SPEEDO ERROR

Indicated	Actual
10 mph	14.0
20 mph	22.8
30 mph	31.8
40 mph	41.5
50 mph	50.8
60 mph	60.5
70 mph	70.2
80 mph	79.8

FORD CONSUL SEDAN
Acceleration thru the gears

Mph (corrected): 10 20 30 40 50 60 70 80 90 100 110 120
SECONDS: 0 4 8 12 16 20 24 28 32 36 40 44 48
1st 2nd 3rd SS1/4 79
ROAD and TRACK

The little racing car silhouette in yellow indicates that the competitor is using Shell products and makes for easier identification at Shell refuelling points

AN ASSESSMENT OF THE EFFECTS OF THE ALPINE RALLY ON A FORD ZEPHYR

Survival of the Fittest

STILL covered with Yugoslavia's clinging white dust, and bearing a sign or two of minor impacts with the rocky walls of the countless passes up which it stormed during the Alpine Rally, the Anne Hall-Yvonne Jackson Ford Zephyr recently came into *The Autocar's* hands. Its female crew, you will remember, put the large car through its six-day ordeal to some effect, finishing 26th overall, and coming second to Nancy Mitchell (M.G.) for the *Coupe des Dames*.

"Why," you may ask, "did you not sample one of the outstandingly successful Triumph TR3s?" The answer is that the Triumph is an acknowledged competition car built for such work, and was appraised recently in its role as high-speed onlooker at the Rally, on which it carried members of *The Autocar* reporting team; its achievements as competitor need no reminder. The Zephyr is a family saloon, of admittedly good performance and handling, but by no means intended primarily to compete in the toughest of modern long-distance rallies. The effects of such an ordeal on a slightly more "bread-and-butter" car are significant and, with respect to the Triumph, I think, of wider interest.

The Alpine Rally was not 513GHK's competition debut. Earlier this year it had competed as a completely standard car in the Tulip Rally in the hands of V. Preston, finishing third in the 2,000 to 2,600 c.c. class for series production cars. Now, as a result of permitted modifications—more carburettors, new manifold, higher compression ratio—of the type which any enthusiast could apply, it is a Special Series touring car.

As sampled by *The Autocar* it had not been touched in any way since it completed the Alpine at Marseilles and was driven back to this country. Speedometer mileage was 16,000. Brakes and clutch were unadjusted; apart from checking the oil level, the engine had not been serviced—the heavy coating of dust, turned into mud here and there by an oil seep, remained as an undisturbed blanket.

Externally, certain deviations from standard were obvious. The steel bonnet and boot tops had been replaced by aluminium pressings in the interests of weight reduction; for the same reason the glass of the side windows had been replaced by Perspex panels. Separate, and extremely comfortable "armchair" seats took the place of the single, three-abreast bench front seat (which has been criticized for allowing the driver to slide about on corners), and held the driver and passenger firmly in place during fast travel on the twistiest of roads. Tailored specifically for the smaller proportions of the car's female crew, the driving position was not ideal for a six-footer, but the increased seat height gave one a tremendous command of the situation and a greatly improved view of the road ahead.

In the space gained between the two seats was a leather-covered box, recessed to hold Thermos flasks, with a compartment for such odds and ends as a plastic bag for a leather and sponge, a torch, cigarettes, chocolate and so on. On the tunnel above the gearbox were two compressed air cylinders for tyre inflation. Special equipment on the dashboard included such things as navigational clocks, and a Halda average speed indicator.

It was noteworthy, on a 600-mile journey undertaken in the car, how extremely useful all this rally equipment can be for normal long-distance motoring; anyone who has frequently to cover great distances would do well to study the equipment built into such cars. The Halda is not a necessity, but it adds greatly to the interest, and one knows at a glance whether one is ahead or astern of a predetermined average speed. Apart from the special navigational equipment, which was grouped mainly in front of the passenger, the main controls and instruments were as on the normal production cars—which, of course, basically this is.

When considering the work carried out on the engine of the car it is necessary to bear in mind that, in production form, the 2.6-litre unit is not, as yet, anywhere near full development. In the interests of reliability and long life it has, in fact, been de-tuned for production purposes. It is particularly interesting, therefore, to see how well the unit stood up to extremely severe conditions when the output had been increased considerably. An indication of the fact that it is

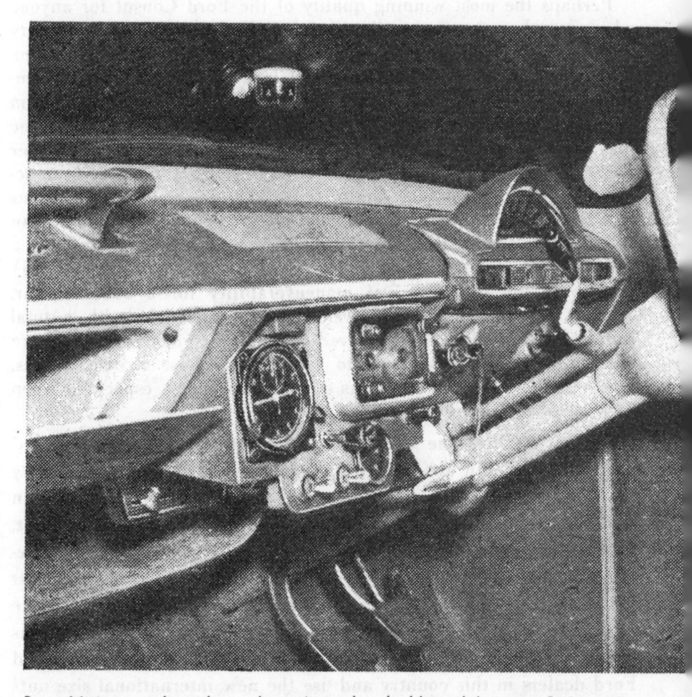

By addition, rather than alteration, the dashboard layout of the rally car provides just about all the information that could be required. Standard Zephyr instruments occupy the right half, and special time-keeping instruments the left, in front of the passenger. The Halda average speed indicator is mounted towards the centre of the dashboard

not a super-tuned engine is the fact that it remains flexible at all speeds and, for example, it will pull evenly between 12 and 100 m.p.h. in top gear.

For most of the 2,500 miles of the rally the engine had been kept at peak power and high revs for long periods, and the car was driven virtually flat-out up pass after pass. The conditions under which a great deal of the mileage was covered were so bad that one team of works-entered cars retired with clogged filters in the dust clouds of Yugoslavia. A total of, I suppose, some 4,500 to 5,000 miles had been covered by the car—including the run to the start, the rally itself, and the return journey—and the engine was still crisp, extremely potent, and without an untoward sound—even from the valve gear.

On opening the bonnet the main, obvious difference from standard lay in the fact that three Zenith carburettors were fitted. The throttle linkage not only looked awkward, but was geared in such a way that the initial, small movement of the pedal opened the throttles far too quickly. This made traffic driving an embarrassment; one was either trickling along on a tickover or, at a touch of the pedal, leaping towards the stern of the next ahead. This was the only adverse criticism

Travel stained, slightly scarred and still wearing its Alpine Rally number—520—the Zephyr is a large car to conduct along endless narrow Alpine passes

of an altogether desirable car, and it concerned a simple shortcoming, easily remedied.

The exhaust and inlet valve diameters had been increased, and the compression ratio raised to 8.5 to 1 by planing the standard cylinder head. Finally, a special easy-flow exhaust system had been fitted, with a Servais silencer. The result of these engine modifications, in conjunction with standard rear axle and gear box ratios, was to give the car a maximum speed in the region of 105 m.p.h. Even with this formidable performance, fuel consumption worked out, over 600 miles, at a best figure of 21.5 and a worst of slightly under 20 m.p.g.

Apart from these performance-seeking modifications, an electric fuel pump had been fitted additionally to the standard mechanical pump, and flexible plastic fuel pipes were used throughout. Ducts had been cut beside the radiator to direct air to the fuel pump. Spring dampers had been fitted giving a 25 per cent increase in hardness of ride; the rear springs were as fitted to the export models, having an extra leaf. All these modifications were per-

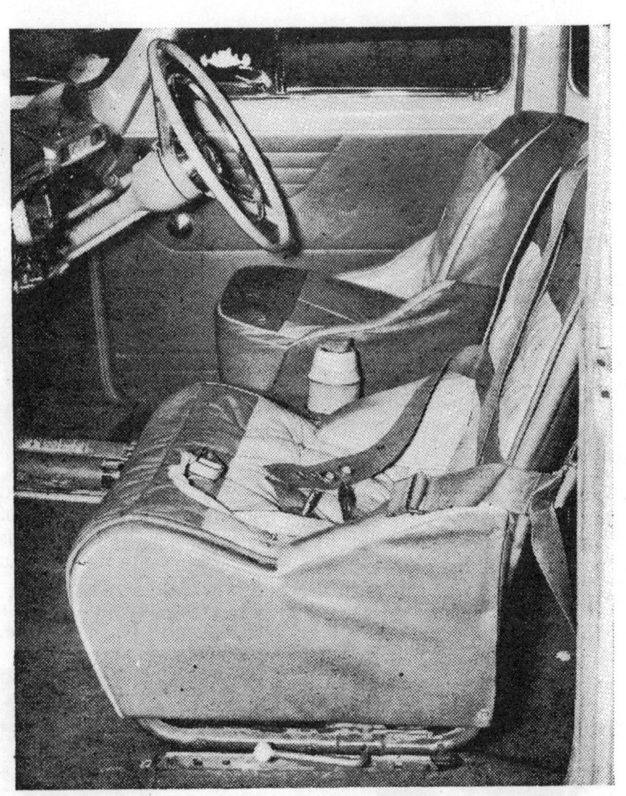

Rally seats which, although they restrict the front capacity to two, are ideal for fast, long-distance travel. A harness was provided for the off-duty driver to prevent her being thrown about on twisty sections

mitted by the Alpine regulations for the Modified Touring class.

During the event the car developed only one defect, which cost it many of the 510 marks lost—a faulty coil. At the departure from one of the stages the engine refused to start and the car was pushed out of the control. Before the start of the Zagreb flying 1,500 metres test a new condenser and a new set of plugs were fitted. Misfiring continued during the run, yet the car managed to exceed the minimum speed for its class; the other Ford Zephyrs exceeded 100 m.p.h.

Later in the rally, the trouble became worse; if the engine was stopped in the mountains when hot, it would not restart; on one occasion the trouble caused a loss of 40 minutes' valuable time. Eventually the trouble was traced to the intermittent operation of the coil at high temperatures. A new coil was fitted and the car ran perfectly. It beat the times set up by the all-male-crewed Zephyrs on the two timed climbs.

The only real effect of the car's strenuous journeyings was found in the clutch, which needed coaxing to avoid slip when accelerating from low speeds in the indirect gears. For this reason it was not possible to obtain any acceleration figures—which should have been most impressive. Brakes had been relined as a precaution on all the Zephyrs during the second halt at Cortina—that is, after the Marseilles to Cortina section, the Circuit of the Dolomites, and the tough section into Yugoslavia and back. The remaining half of the rally, which included Cortina to Megève, and Megève down to Marseilles—and an endless succession of the worst passes in Europe—together with the return drive to England, had been completed on the present set of linings. Adjustment of brakes on the Rally was, of course, frequent. Though there was plenty of pedal travel—through lack of recent adjustment—the brakes were still good, and pulled the car up in a straight line even when the driver's hands were off the wheel. Surprisingly light pedal pressures were required for the average main-road slow-down from high speeds, which was unexpected in view of the fact that hard, non-fade Mintex M linings were used.

The stiffened suspension produced steady and level cornering, the car adopting a slight roll on entering which remained constant throughout the corner. The car's hand-

Survival of the Fittest . . .

Dust reduces the majority of the under-bonnet scene to an all-over grey. The three Zenith carburettors can be seen, and the big air-cleaner which did much to keep the car going in the dust clouds of Yugoslavia. The somewhat complicated throttle linkage will be noticed to the left centre of the picture

ling on corners was of the type one expects from true-bred sporting machines; on the few occasions when owners of much more potent vehicles decided to give battle, it was found that the Zephyr could more than hold its own on open, flat-out bends.

There were few rattles in the bodywork, despite the car's having travelled fast on some of the worst roads in Europe.

Lack of attention accounted for some creaks and groans in the rear suspension, particularly during acceleration when spring wind-up occurred. Largely because of the difficulty in opening the throttles gently, there was a slight bump as the reversal of loading occurred in the transmission on accelerating away from the overrun. It must be remembered, however, that the car had received no attention whatever for many miles. There are few of these faults that a thorough servicing would not have cured.

The steering, slightly heavier than on the standard car—partly the result of Michelin X tyres and partly the complete absence of lubricant—was accurate, positive, and had very little idle movement. It was an easy matter to place the car exactly where one wanted, and to guide it through narrow gaps.

Summing up, the car was still taut, safe and immensely exhilarating. It represented with its enhanced performance and its rally equipment, just about the ideal for fast, long-distance travel and it gobbled up the miles like a pair of seven-league boots.

It demonstrated convincingly that the British everyday saloon, in standard form, can hold its own on rough going without the need for major chassis modifications—and that, suitably tuned, it can be made to give the performance of a died-in-the-wool sports car. Such a showing—by this and the other British cars that did so well in the Alpine—should go far to boost the sales curve. P. G.

Yvonne Jackson (left) and Anne Hall still manage to look very smart (unlike some of the male crews) and feminine at the end of the Alpine

CONTINUED FROM PAGE 15

habit of waving like a lily of the field. On the other hand the change is not as smooth as some and there is no reverse-catch. The lever is on the left side of the wheel. In engaging bottom gear from rest one is apt at times " to hit the cogs."

The headlamps give good, if not particularly penetrating, converging beams, and can cope well with normal fog when dipped. The foot dipper is well placed. The rear-view mirror is adequate, the central roof lamp and the instrument lighting excellent. Suction wipers are fitted, but their wiping speed is adjustable and they functioned well; their control knob pulled off but could be replaced. The indicators had a slight tendency to stick and one rear lamp bulb failed.

The body is completely waterproof and draught-free and was free from all save one major rattle in the near-side front door. The fuel filler, placed horizontally, is a cause of slight anxiety to those who like to refuel from a can.

Easy cruising at speeds commendably high for a car of this kind and the high degree of pleasure afforded by the taut steering and good roadholding, explain the growing popularity of this car, and its sister, the $1\frac{1}{2}$-litre four-cylinder Consul, on British roads. Ford has every reason to feel proud, and sure of the continued success of a saloon so well suited to the needs of so many motorists, especially bearing in mind its competitive basic price of £532. Indeed, remembering how well the Ford V8, Pilot, Anglia and Prefect served the people, backed by world-wide service, it can be said, of the Consul and Zephyr-Six, that " Dagenham has done it again ! "—W. B.

THE FORD ZEPHYR-SIX SALOON

Engine : Six cylinders, 79.37 by 76.2 mm., 2,262 c.c., push-rod o.h.v.; 6.8 to 1 compression ratio; 68 b.h.p. at 4,000 r.p.m.
Gear ratios : 1st, 12.62 to 1; 2nd, 7.29 to 1; top, 4.44 to 1.
Tyres : 6.40–13 Goodyear on bolt-on steel disc wheels.
Weight : 23 cwt., without occupants but ready for the road with one gallon of fuel.
Steering ratio : $2\frac{1}{2}$ turns, lock-to-lock.
Fuel capacity : Nine gallons. Range approx. 225 miles.
Wheelbase : 8 ft. 8 in.
Track : Front, 4 ft. 2 in. Rear, 4 ft. 1 in.
Overall dimensions : 14 ft. $3\frac{3}{4}$ in. by 5 ft. 4 in. by 5 ft. $0\frac{3}{4}$ in. (high).
Price : £532 (£829 1s. 1d. with p.t.).

PERFORMANCE DATA

Speeds in gears :
1st ... 28 m.p.h. Top ... 80 m.p.h.
2nd ... 50 m.p.h.
Acceleration through gears :
0-50 m.p.h. in 13 sec. s.s. $\frac{1}{4}$-mile in 22 sec.
0-60 „ „ 19 „
Makers : Ford Motor Co., Ltd., Dagenham, Essex.

The Autocar ROAD TESTS 1652

Ford Zephyr

ESTATE CAR

ALTHOUGH the main object of the Road Test of the Ford Zephyr estate car was to discover how this model compares with the normal saloon in comfort, performance, economy and handling, the opportunity to renew acquaintance with a Zephyr was anticipated with pleasure.

When tested over a year ago (*The Autocar*, 13 April 1956) the redesigned Ford Zephyr had then only recently appeared. The assessment of the saloon was that it was one of the best and most encouraging British cars in large-scale production that *The Autocar* had tested since the war. The impressions gained during the current test give no cause to amend that appraisal. Many of the comments arising from that first test and from a subsequent test of a Zodiac with automatic transmission (*The Autocar*, 8 February 1957) apply equally to this estate car version.

In mechanical details the Zephyr estate car, which is converted from the standard saloon by E. D. Abbott, Ltd., of Farnham, is identical except for the use of stiffer rear springs, and larger section tyres—6.70in replacing 6.40in on 13in rims. The rearward extension of the body above what would be the luggage locker of the saloon is of welded steel construction, and provided with large windows at the sides and rear. A single, side-hinged loading door is fitted. Because of the pronounced forward lean of the rear panel, there is some upward movement of the door when it is opened, but a check arm prevents the door from opening beyond the width of the car and also holds it automatically in the open position. The corners of this door seemed unnecessarily sharp.

The opening for loading is wide but rather shallow, and there is a ledge which prevents luggage from being slid directly on to the loading platform. This platform is, in effect, an extra wooden floor above the boot platform of the normal saloon, and access to the spare wheel below it is through a trap-door in the floor. Tools and wheel-changing kit are housed with the wheel. Alternatively the model can be had without the extra floor, when the spare wheel is mounted upright in a well and the existing boot floor is used for loads.

Front and rear seats are similar to those of the saloon, except that the rear squab and cushion may be easily moved to increase luggage-carrying space. The rear seat squab is released by moving a draw-bolt at each side, and it can then be lifted out and placed behind the front seat in a vertical position, with a strap suspended from each side of the front squab to hold it upright. The rear seat cushion then becomes a continuation of the luggage platform or, alternatively, the cushion may be removed completely to increase the space.

This very simple arrangement for moving the rear seat has the disadvantage that with the rear cushion in its usual position, it is unprotected from loads which may be carried on it. This matters less when tough plastic upholstery covering is used as on the model tested, but the very light colours chosen—in this case near-white and orange—could be marked very easily. An advantage, however, is that cushion and squab can readily be removed for use outside the car, for example, when picnicking.

Again, the mat for the floor of the luggage compartment is in very light-coloured but durable rubber material, having a raised diamond pattern. It is not easily marked by heavy objects, but quickly traps dirt. The mat, held down by press fasteners, can be readily removed, but the hinges of the trap-door stand slightly proud below it and have to be avoided when loads are slid into position. The mat can be cleaned by brushing with soap and water. The back of the rear seat squab and the wheel arches have the same covering to protect them. There is a durable body lining below window level and around the frames of the side windows. A light in the rear quarter, to the left of the door, makes loading at night an easy matter. A small point, but one much appreciated, is the padding around all door frames, which softens the blow if the edge is struck accidentally.

The rearward extension, above waist level, gives an illusion of greater length compared with the normal saloon. The roof line is slightly lower above the luggage space. The single loading door has a large window and concealed hinges. Each over-rider contains a light to illuminate the number plate, behind which is the fuel filler

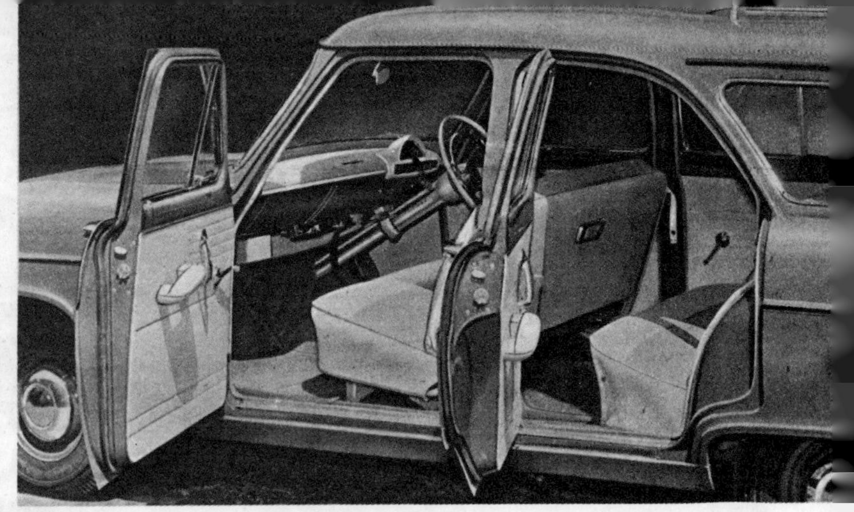

Maximum luggage space is provided when the rear seat squab is placed against the front one. Two straps retain it in position. An arm checks the loading door and holds it open. Forward of the rear seat, the interior is identical with the standard saloon. There is ample knee and leg room for rear seat passengers with the front seat in its rearmost position of adjustment

Ford Zephyr Estate Car . . .

The weight of the estate car is some $1\frac{1}{8}$ cwt more than that of the saloon—a small sacrifice when the greatly increased carrying capacity of this body is taken into account.

It had been expected that acceleration would suffer, because of the increase in weight and the slightly higher gearing provided by the larger section tyres, and also that the rectangular shape of the rear of the car would increase drag and lower the maximum speed; in fact the figures approached very closely to those of the saloon. The standing start quarter-mile was covered in 20.9 sec compared with 20.5 sec, and although the 0-30 m.p.h. figure for acceleration through the gears was virtually the same, there was a progressive increase in the time to reach the higher speeds, the figure for 0-80 m.p.h. being 4.4 sec greater.

Maximum speeds in the gears were higher, but mean maximum speed in direct top (a Borg-Warner overdrive was fitted to the car tested) was 1 m.p.h. slower—a discrepancy which might well be found between two saloon models.

Of greater importance, probably, to the owner of an estate car is the fuel consumption with different conditions of load. The overall figure of 25 m.p.g., taken over a distance of more than 600 miles of varied motoring is, in fact, slightly better than the figure obtained during the road test of the standard saloon. Hard driving with the car heavily laden, reduced this figure by only about 3 m.p.g.—a very creditable performance. When the car was driven easily and maximum speed was restricted to about 55 m.p.h., but without using any special trick methods, the very good figure of 33 m.p.g. was obtained with two up and a quantity of luggage, on a single journey of 80 miles. Throughout the test the overdrive (which is optionally extra) was used, and judging by previous experience of the Zephyr it is probable that some 2 to 3 m.p.g. improvement was attributable to the overdrive.

Two of the outstandingly good points of the Zephyr are engine smoothness and handling qualities which combine to give this six-seater family car a very strong position in its particular price category. High average speeds on long journeys are possible without fatigue of the occupants, for little engine or road noise is heard inside the car, which can be cornered briskly with complete confidence. The material used for the seat covers effectively prevents the passengers from sliding. Steering which is light—even at manœuvring speeds—and accurate, with strong self-centring action, exhibits no shortcomings when the car is near the limit of lateral adhesion on wet roads. Wheelspin is easily induced when starting on slippery surfaces. There is, perhaps, slightly less comfort, particularly for rear-seat passengers because of the stiffer rear suspension on the estate car, but when a load is taken aboard this slight harshness disappears. Damping is excellent and fully up to its task, even when maximum weight is carried over rough going.

Brakes, although of ample power with low pedal pressures, and free from fade in normal road use, exhibited some roughness when applied from high speed. During an emergency stop on a dry surface, all four wheels were locked, and the car remained on a straight course.

Estate cars in general are subject to big changes in weight distribution between front and rear wheels with varying loads, and it would be unfair to hope that handling qualities would remain unchanged. A reasonable owner of an estate car would not expect, when his car was carrying the maximum load, to take the liberties which would be permissible when driving a normal saloon. However, it is better that he should know what to expect if circumstances cause him to be unduly enterprising, and we can reassure the prospective owner of a Zephyr estate car that the handling qualities are less affected by load than are those of most vehicles of this class.

A considerable distance was covered with a 6 cwt burden on the platform above the rear axle, and two persons in the car. Rear tyre pressures were raised by 4 lb sq in. Although there was greater roll on corners, there was no suggestion of increased lightness or lack of precision in the steering, and the car was only slightly less responsive to quick changes of direction. The effect on acceleration was hard to detect, braking power was sufficient to cope with the load, even when the overdrive permitted the free-wheeling below about 30 m.p.h., which the Borg-Warner unit provides. Steep hills were climbed without the impression that more power would be useful, and it was felt that generally the car, thus laden, was robust

Among its many functions the Zephyr estate car can be put to good use on the farm. The built-in roof rail, rubber covered slats and strap eyes are standard equipment

enough to withstand the very severe conditions abroad.

The maximum permissible gross weight of this model is 4,000 lb, which means that it is possible to load the car with either passengers or luggage or goods up to about 1,112 lb, assuming the fuel tank to be full. With two aboard nearly 7 cwt may be carried in the rear, the goods capacity decreasing, of course, if more passengers are carried, until with six up, 1 cwt of luggage will not overload the car.

Although a wide angle of view is given by the driving mirror, vision some distance behind the car is cut off by the upper edge of the rear window frame. When the tail is depressed by a heavy load, the range of vision is further restricted. An external mirror is essential for an estate car in which loads may obscure the backward view from within the car. Reversing is simplified to some extent by the proximity of the rear window to the bumper, though the wing extremities are hidden.

In this model the appeal of extra space and carrying capacity is added to the recognized good features of the Zephyr. Furthermore, this is achieved with such a small performance penalty as to be barely perceptible, and fuel consumption shows a slight improvement. Although physically capable of the rougher work sometimes expected of an estate car, the light and elegant interior of the example tested was more suited to the carriage of large quantities of normal luggage. The availability, as an alternative, of a more utilitarian trim for the load space might well widen the already large appeal that this model possesses.

FORD ZEPHYR ESTATE CAR WITH OVERDRIVE

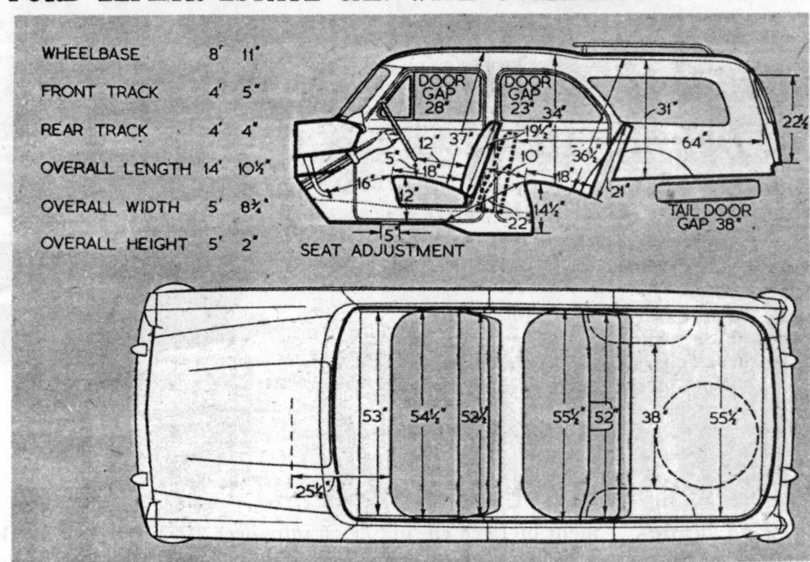

Measurements in these ¼in to 1ft scale body diagrams are taken with the driving seat in the central position of fore and aft adjustment and with the seat cushions uncompressed

PERFORMANCE

ACCELERATION: from constant speeds.

Speed Range, Gear Ratios and Time in sec.

M.P.H.	*2.83 to 1	3.90 to 1	*4.49 to 1	6.40 to 1	*7.76 to 1	11.08 to 1
10—30	—	8.9	—	5.3	—	3.8
20—40	—	8.9	—	5.7	—	—
30—50	14.9	9.3	8.6	6.5	—	—
40—60	17.3	11.2	9.7	9.9	—	—
50—70	23.8	14.3	13.8	—	—	—

* Overdrive

From rest through gears to:

M.P.H.	sec
30	4.8
50	12.6
60	18.7
70	27.3
80	40.2

Standing quarter mile, 20.9 sec.

SPEEDS ON GEARS:

Gear		M.P.H. (normal and max.)	K.P.H. (normal and max.)
O.D. Top	(best)	80	128.7
Top	(mean)	83	133.6
	(best)	84	135.2
O.D. 2nd	..	75—82	112.7—132.0
2nd	.. ..	50—61	80.5— 98.2
O.D. 1st	.. ..	43—50	69.2— 80.5
1st ..	.. ..	26—36	41.8— 57.9

TRACTIVE RESISTANCE: 34 lb per ton at 10 M.P.H.

TRACTIVE EFFORT:

	Pull (lb per ton)	Equivalent Gradient
Top	249	1 in 6.3
Second	395	1 in 5.6

BRAKES (in neutral at 30 m.p.h.):

Efficiency	Pedal Pressure (lb)
32 per cent	25
68 per cent	50
82 per cent	75

FUEL CONSUMPTION:
25 m.p.g. overall for 682 miles (11.3 litres per 100 km).
Approximate normal range 22-33 m.p.g. (12.8-8.6 litres per 100 km).
Fuel, Premium grade.

WEATHER: Dry, light cross wind.
Air temperature 72 deg F.
Acceleration figures are the means of several runs in opposite directions.
Tractive effort and resistance obtained by Tapley meter.

SPEEDOMETER CORRECTION: M.P.H.

Car speedometer ..	10	20	30	40	50	60	70	80
True speed ..	9	18	28.5	39	49	59	69	77

DATA

PRICE (basic), with saloon body, **£817 10s.**
British purchase tax, **£410 2s.**
Total (in Great Britain), **£1,227 12s.**
Extras: Heater £11, plus £5 10s purchase tax.

ENGINE: Capacity: 2,553 c.c. (155.8 cu in).
Number of cylinders: 6.
Bore and stroke: 82.55 × 79.5 mm (3.25 × 3.13in).
Valve gear: o.h.v., pushrods.
Compression ratio: 7.8 to 1.
B.H.P.: 90 (gross), 85 (nett) at 4,400 r.p.m. (B.H.P. per ton laden 63.7).
Torque: 137 lb ft gross at 2,000 r.p.m.
M.P.H. per 1,000 r.p.m. on top gear, 19.16.
M.P.H. per 1,000 r.p.m. on overdrive, 27.37.

WEIGHT: (with 5 gals fuel), 25⅓ cwt (2,838 lb).
Weight distribution (per cent): F, 54; R, 46.
Laden as tested: 28⅓ cwt (3,174 lb).
Lb per c.c. (laden): 1.24.

BRAKES: Type: Girling.
Method of operation: Hydraulic.
Drum dimensions: F, 9in diameter; 2¼in wide. R, 9in diameter; 1¾in wide.
Lining area: F, 86.5 sq in. R, 60.5 sq in (104 sq in per ton laden).

TYRES: 6.70—13in.
Pressures (lb per sq-in): F, 26; R, 26 (normal).

TANK CAPACITY: 10½ Imperial gallons.
Oil sump, 7 pints.
Cooling system, 22 pints (plus 1 pint if heater is fitted).

TURNING CIRCLE: 36ft (L and R).
Steering wheel turns (lock to lock): 3.

DIMENSIONS: Wheelbase: 8ft 11in.
Track: F, 4ft 5in; R, 4ft 4in.
Length (overall): 14ft 10½in.
Height: 5ft 2in.
Width: 5ft 8¾in.
Ground clearance: 6.8in.
Frontal area: 22 sq ft (approximately).

ELECTRICAL SYSTEM: 12-volt; 57 ampère-hour battery.
Head lights: Double dip; 42-36 watt bulbs.

SUSPENSION: Front, Independent coil springs; anti-roll bar. Rear, Semi-elliptic leaf springs.

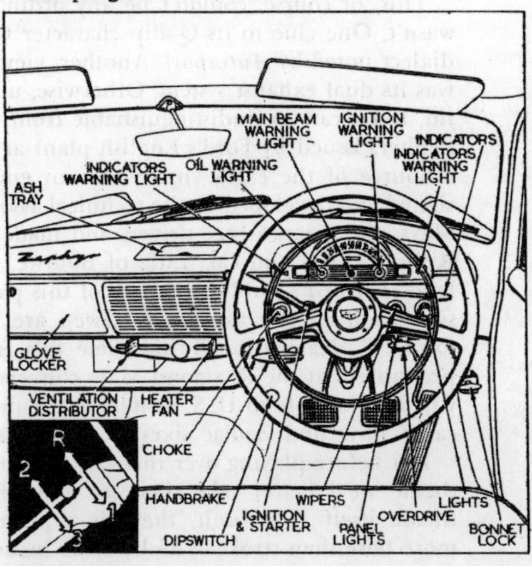

GALE

A simple bolt-on kit can make Ford's Zephyr go like its big brothers from Detroit. All this and economy too - what more can you ask?

Mays-treated test car differed from standard Mays setup by having the larger (2H6) carbs delivering 11 hp extra. Stock R.M. layout uses SU (2H4) carbs.

By DENNIS MAY

KEN WHARTON's last race in his native land, before crashing fatally at the wheel of a Ferrari at Ardmore, New Zealand, early in January, made a sharp impression on intellects that don't impress too easily. This was a ten lap event for stock and modified sedans over the weavey Oulton Park circuit, 2¾ miles around, in Cheshire, England. Wharton, vacationing from the FI and sports-racing cars that had latterly occupied most of his time, drove a Ford Zephyr and won any way he liked. *Autocar* described the Zep as "incredibly fast . . . and very stable," *Autosport* adding that it "handled and sounded like a Formula I car."

Wharton's race average was 69.44 mph, which, considering nobody gave him a run for his money, compared favorably with Ivor Bueb's 74.39 per in the concurrent *Gran Turismo* contest on the fastest 300SL Merc in the country.

This, of course, couldn't be any ordinary Zephyr, and it wasn't. One clue to its Q-ship character was that Formula I dialect noted by *Autosport*. Another, viewing it from astern, was its dual exhaust system. Otherwise, unless you lifted the lid, Ken's car was indistinguishable from the commonalty of Zephyrs issued by Ford's English plant at Dagenham, Essex.

Source of the extra virility was an engine makeover designed and developed by the technical associates of Raymond Mays, the veteran race driver, and made and marketed by Rubery Owen and Co., Ltd., of Bourne, Lincolnshire, England. For *SCI*'s public, the facts of this pack's parentage are significant insofar as Rubery Owen are now on the scent for U. S. dealerships. By the time this issue is on sale it's probable that the Raymond Mays conversion, as it is called, will be available to U. S. owners of the mechanically identical Zephyr and Zodiac sixes from domestic sources.

But before playing over this variation on the Ford engine theme we'd better take time for a brief rendition of the theme itself. This unit, then, is a pushrod ohv six with more bore than stroke—3.25 by 3.125 inches—and a displacement of 155.8 cu. in., i.e., a mite over 2½ litres. The crank-case-cum-cylinder carcase is an iron casting, and so is the detachable head. The valves are in line but inclined at 14 deg. to the vertical. Inlet and exhaust ports, the former siamised and the latter separate, are both on the left side of the head. Carburetion is by a single downdraft Zenith 34WIA and spent gas makes a rather uneasy exit into an exhaust pipe pierced with holes corresponding to the ports and clamped directly to the head (there is no manifold in the ordinary sense). The crankshaft is cast iron and runs in five bearings; journal and crankpin diameters are 2⅜ and 2 inches respectively.

With the standard compression ratio of 7.8 to 1, this powerplant develops 86 horsepower at 4200 rpm and a maximum torque of 136 lb/ft at 2000 per minute.

Kernel of the R.M. conversion is an aluminum alloy cylinder head in a material known as DTD424, the same metal that Jaguar used on all the XK variants and derivatives. Except insofar as the depth of the wedge shaped combustion chambers is reduced to give a higher compression ratio, the interior form of this head is similar to Ford's own; direction of squish is towards the 14 millimeter spark plugs, which are on the remote side from the ports and set at a slightly downward slant. The porting, however, breaks right away from Dagenham's design. Inlet tracts have a semidowndraft inclination of 45 degrees, and this, in conjunction with valves at the decreased angle of 4 deg. to the perpendicular, makes for very free flowing passages.

The inlet manifold too is an alloy casting and mounts dual SU carburetors at a slope that continues the line of the ports. The manifold has a center dividing wall with a connecting drillway to balance the ration of charge between the two groups of cylinders.

In the offtake department, Ford's rudimentary plumbing

FORCE FOR THE ZEPHYR

Visibility from Zephyr's driving seat is equal to current Detroit standard, better than British. Wheel angle is restful, gives good control at high speed.

Wide compartment gives good accessibility to plugs, valve gear etc. Kit is slated to be marketed here in very near future.

Kit includes high compression, aluminum alloy head, dual semidowndraft SU carbs, header system, oversize valves, tube push rods.

FORD ZEPHYR WITH RAYMOND MAYS CONVERSION

PERFORMANCE

TOP SPEED:

	R. M. Conversion	Stock
Two-way average	102.2 mph	86.0
Fastest one-way run	103.4 mph	—

ACCELERATION:

From zero to	R. M. Conversion	Stock
30 mph	4.0 secs.	—
40 mph	5.1 secs.	—
50 mph	8.6 secs.	12.3
60 mph	10.3 secs.	17.9
70 mph	15.0 secs.	25.4
80 mph	19.4 secs.	35.8
90 mph	23.3 secs.	

	R. M. Conversion	Stock
Standing ¼ mile	18.4 secs.	20.5 secs.
Speed at end of quarter	78 mph	63 mph

SPEED RANGES IN GEARS:

I	0 to 32 mph
II	7 to 60 mph
III	10 to 103.4 mph

SPEEDOMETER CORRECTION:

Indicated	Actual
30	27
40	36
50	46
60	55
70	67
80	75
90	85
100	95

FUEL CONSUMPTION:

Hard driving	15.8 miles per U.S. gallon
Average driving (under 60 mph)	See text.

BRAKING EFFICIENCY:

(10 successive emergency stops from 60 mph, just short of locking wheels.)

1st stop	70
2nd stop	71
3rd stop	72
4th stop	71
5th stop	70
6th stop	69
7th stop	64
8th stop	64
9th stop	66
10th stop	65

Zephyr Conversion

is replaced by two separate cast iron exhaust manifolds, each collecting from three ports and discharging into its own downpipe to a Servais sound-absorbtion silencer. The upper faces of these manifolds are in contact with the bottoms of two square section wells formed in the floor of the intake gallery; functions of the wells are to act both as hotspots and traps for any liquid fuel that may invade the system as a result of over liberal use of the mixture control during warmups.

Both rows of valves are oversize and of nonstandard material. Head diameter of the inlets is 1.65 inches, exhausts 1.29. Steel for the former is to specification EN110, the latter EN59, a highly durable and heat resistant silicon chrome alloy. Port sections too are increased, and the 2H4 carburetors normally fitted have 1½ inch throats. A single spring per valve is retained but the strength is in excess of the Ford pattern.

The whole bottom end, including the camshaft, stays unmodified, but reciprocating weight in the valve gear is cut by replacing the solid pushrods with tubular ones. By compensatory design work the original rocker geometry is maintained in spite of the altered valve angle.

Experiment and development is in constant progress at Bourne, and already there are several embellishments on the basic conversion that can be specified. In the form outlined above, however, and with the regular compression ratio of 8.75 to 1, the Mays-treated Zephyr engine turns 127 bhp at 4750 rpm. Maximum torque in this trim is 153 lb/ft at 3000 rpm. Pending the establishment of U. S. agencies, Rubery Owen welcomes direct enquiries at Bourne. Pricewise, all that can be said at this writing is that in Britain the pack sells for the sterling equivalent of around $400.00. There is an additional installing charge in the U.K. approximating to $35.00 but detailed fitting instructions are supplied free and this isn't a job to daunt any average competent amateur mechanic.

Of the various measures used to boost output beyond the 127 horsepower level, all except one, viz., an optional compression ratio of 9.2 to 1, incur a surcharge of some kind. This high ratio, claimed to yield an extra 5 bhp, adds nothing to the price but calls for 100 octane gas. Wharton's car at Oulton Park, which was the makers' prototype, entered to whet public interest in advance of production, was running at 9.2 to 1 and had its exhaust pipes duplicated throughout their length, instead of only as far as the silencer. Figures to show the specific bhp value of dual flues are not available, but this system certainly talks your ear off at anything over half throttle.

Other price-inflating options are oversize carburetors (type 2H6 SUs with 1¾ inch throats), said to raise the power to 138 at 5000 rpm; and full race cams of evidently brutal characteristics—they contribute a further 18 horsepower, assert R.O.

As evidence of Rubery Owen's determination to persevere until a Zephyr engine finally blows up in their faces, we saw one of these mills on the test bench with three enormous dual-choke Weber carbs fitted. Assuming it also had the hot camshaft (the exercise was purely experimental and inquisitions were discouraged), this tinderbox might ultimately be expected to peak at well over 160 bhp.

Although obviously a radical improvement on Ford's conception of exhaust manifolding, the two cast iron collectors of the Mays conversion still fall short of racing practice as regards freedom of gas flow. So, in the course of our snoop around the test shop at Folkingham airfield, the Bourne outpost where most of Rubery Owen's proving work is done, it was interesting to come upon an authentic banana cluster exhaust system in Zephyr measurements. This comprised two fabricated assemblies instead of castings and was a model of delectably merging curvatures. Presently and in the forseeable future, it is not for sale.

Although relatively inexpensive, the Mays conversion hits a high standard of workmanship and finish. That isn't surprising when you consider that the man primarily responsible for it has for many years devoted nine tenths of his time to pure race projects, mostly conducted in a cost-no-object climate. He is Peter Berthon, head of the team that designed and is still developing Britain's fastest Formula I car, the 2½ litre BRM. (Experience fortunately suggests that the warmed up Ford doesn't share the BRM's mercurial temperament). Alfred Owen, millionaire head of the Rubery Owen industrial group, is also of course the cashbox of the Owen Racing Organization, which owns and operates the Grand Prix BRMs.

Using a converted Zephyr supplied by R.O., we have made a first-hand evaluation of this enterprising metamorphosis. The performance figures tell their own story, and a remarkable one it is. They nevertheless lose most of their significance unless related to statistics for the stock Zephyr, which has not been roadtested by *SCI*.

Startling as these figures seem, they come as no surprise to a driver who has tried both stock and converted Fords. The Raymond Mays treatment literally transforms the car, giving it a bounding exuberence that lifts it way out of the ruck of English sedans in the 2 to 3 liters bracket.

The Mays setup on the test car differed from standard in having the larger (2H6) carburetors, which are reckoned to be worth 11 bhp extra. Rubery Owen states, however, that this gain comes at the top end of the power range, the torque low down being if anything inferior. This seems to be corroborated by the facts that our maximum speed was up by about two mph on a contemporary's findings with the smaller SUs, whereas all our acceleration figures were down by assorted nuances.

The car on test was also fitted with Borg Warner semiautomatic overdrive, which is one of Dagenham's transmission options. This particular B.W. application is engineered to make its up-shifts at 31 mph, whether out of low, second or high; so, as 31 mph is exceedable in low, it follows that six ratios are on call, which in theory should lend the thing wings in timed acceleration tests. In practice, though, it doesn't work that way, due to the measured pause punctuating the shifts. After timing runs with and without benefit of Borg Warner we came to the conclusion that two mauling manual changes, and just that, incurred a smaller penalty than the alternative, in spite of the wide spacings involved.

Normal high on the test car was 4.1 to 1, giving 17.6 mph per 1000 rpm. Corresponding speed on overdrive high was 25.2 mph, from which it will be deduced, and rightly, (a) that overdrive adds a lot to restfulness when cruising fast and far, and (b) that maximum speed in this ratio, except with a following wind or going downhill, falls short of direct top maximum.

There is a connection, incidentally, between the B.W. overdrive and the retention of single valve springs in this conversion. Although the Mays job, with big carbs as tested, reaches its power peak 800 rpm further up the scale than a stock Zephyr, the former makes it plain that it would rev faster still, and like it, if its valve springs would let it. At or around a corrected 60 mph in second, for instance, the valve gear quite suddenly starts yammering like a ruptured fairy. We commented on this to a Rubery Owen luminary and were told that the decision to stick to single springs had been partly dictated by isolated cases of a hamfooted driver inadvertently going beyond the full throttle travel and kicking himself out of overdrive when making say 80 per in o/d second, or some unprintable speed downhill in o/d high. When this happened, double springs broke up but single ones didn't. Presumably the range of bounce allowed by single springs is insufficient to cause piston/valve collisions.

Our test itinerary being much curtailed by fuel rationing, it wasn't possible to get separate sets of mpg figures for hard *and* average driving over significant mileages: by the time we were through driving hard, it was time to stop driving at all. All we can add to the solitary consumption return shown in the table, therefore, is that Rubery Owen claims that the conversion with big bore carburetors gives approximately the same mpg as the stock Zephyr at equivalent speeds and goes slightly further per gallon when using the 1½ inch carbs.

In every other aspect of behaviour the converted engine is in our opinion either equal to or better than its standard counterpart. It starts instantly, whether cold or when freshly switched off after serial full-bore acceleration bursts; it idles sweetly and regularly, hot or cold; the level of mechanical noise, except when valve yammer is allowed to develop, is moderate—certainly no worse than the iron head job; most important of all, this is, for its displacement, a flexible engine with good pulling power at low to medium turnover. In common with the unmolested version, it gives very easy access to all departments liable to need periodic attention—carburetors, spark plugs, tappets, distributor, etc. The alloy head itself is 28 pounds lighter than the normal one, but the extra weight of such items as the cast iron exhaust manifolds, two carbs and air cleaners instead of one, and so forth, pretty well restores the status quo.

Exhaust noise from the basic Raymond Mays system, using a single tailpipe, is surprisingly subdued, scarcely louder than with Dagenham's own plumbing and far lower output.

Ford of England, in common with Vauxhall, the only other British marque with American affiliations, persist in the un-English practice of omitting a fourth ratio from their gearbox. No doubt a proportion, maybe a majority, of the R.M.-converted Zephyrs and Zodiacs that ultimately go into service in the U. S. will be fitted with Borg Warner overdrive, and thus have six speeds on call. However, having driven the loaned car with its overdrive locked out of action for more than half of our test mileage, we can fairly evaluate the combination of a manual three-speed transmission with the engine characteristics that the Mays recipe produces. On average British roads and in typical British traffic densities, which dictate a more fitful motoring tempo than most U. S. drivers are accustomed to, a three-speeder labors under an inherent handicap—unless its engine has really useful mid-range torque. This the R.M. Zephyr certainly does have, which very largely compensates for the absence of a fourth cog.

The point can be illustrated by comparisons with another British sedan of roughly equal displacement that SCI has roadtested. The 2.4 Jaguar, with four speeds and a handier and more positive acting gearshift, is 2.2 seconds slower than the Ford from zero to 70 mph.

With a gap of 2.3 ratios between second and high in the Ford drive train, it could be forgiven if up-shifts between these stages produced momentary power doldrums. In practice they don't. On the contrary, when the revs are taken to the brink of valve bounce in second, top comes in with a vigor that leaves a brief wheelspin smudge on dry concrete.

The shift itself, of course, is on the steering post, and, as these devices go, it isn't at all bad. Its range of travel is reasonable, and the synchromesh on second and third works well and isn't all the time panting "Hey, wait for me."

For customers who feel that the non-standard performance of the converted Zep calls for non-standard aids to safety and road holding, Rubery Owen offers, at extra cost, harder brake linings and heavier duty Armstrong dampers for the back springs. *SCI's* test car had the latter (which Armstrong has engineered in consultation with R.O. specially for this car), but not the former. As our braking efficiency percentages show the normal linings stood up pretty well to *SCI's* regular gruel of ten hard stops from sixty in rapid succession. Pedal pressures were reasonable but the last few detentions produced unavoidable locking of one rear wheel or the other.

If it is true — which it isn't literally, of course — that the R. M. Zephyr "handles like a Formula I car," much of the credit belongs to the stringent damping at the rear end. It's remarkable how far this single suspension adjunct goes towards imparting sports car feel and cornering characteristics to what is normally a billowy and, by European standards, roll prone family conveyance. Earlier experience on a Zephyr with original pattern shocks had shown that with no center armrest to keep the driver and his neighbor apart on left-hand corners (the car had righthand steering, of course), the latter was apt to become the plaything of transverse *G* and wind up literally in the conductor's lap. This apparently was the price that had to be paid for a very comfortable ride in a straight line and over bad surfaces. The R. M. Zephyr too lacks a center armrest, but it is hardly missed. The supplementary damping puts lateral stability onto an entirely different plane and, by preventing an exaggerated angle of lean on turns, indirectly improves cornering power very markedly.

At any speed within its scope, this Ford holds a true course during straight running, answering the helm in a consistently predictable way, even if the steering linkages do convey a hint of the prevailing modern flaccidity. The same as in unmodified form, there is an acceptable trace of understeer. Three turns of the wheel winch her from lock to lock, which isn't excessive, but the turning diameter of 36 feet is too big for maximum convenience in the tight maneuvers demanded by rally tests. A rally man with the Mays appurtenances under his hood has the answer to that problem, though: he can use his power bonus to steer through the back wheels.

Dennis May

Ford Consul Road Test

(Continued from page 20)

The parking brake is conveniently located under the facia at the left of the steering column and seemed quite efficient for the purpose for which it was designed.

Overall body design is highly functional, with clean and modern lines. It is free from unnecessary embellishment. Wind noise at speed is low and unobtrusive. Road noise and drum is present, as with most of the mono-constructed body-and-chassis cars we have tried, and should be improved.

The interior is tastefully furnished. The raised and hooded instrument panel carries a large half-circle speedometer, with fuel and ampmeter gauges.

The wipers are vacuum-operated. The horn is operated from a press button in the centre of the steering wheel, the most practical method. The trafficator arm switch is above. A full-length parcel shelf runs across the facia. Ashtrays are provided front and rear.

The floor controls are well spaced, with pendulum pedals for clutch and brake and accelerator. There is room off the clutch for the left foot, where it can rest on the dip switch at night.

The seating is not above criticism. The benches are covered in leather cloth and are rather too hard. In addition, the driver finds his bench rather short under the knees, giving a "sitting on the point" effect which is tiring after a long drive. In addition the seat adjustment locking device is not as positive as it could be, particularly on the passengers' side.

The roof is lined with a washable plasticised material, with a centre light.

Headroom throughout is ample, but leg room in the rear is cramped.

For the family man the very large rubber-lined boot is an outstanding feature. The lid is hinged ruggedly at both sides and is spring-loaded. The spare wheel is housed at the right, providing space for luggage.

Ventilation in the interior is provided by quarter-light front windows and a vent ducting air from the front of the car, controlled by a pull-out knob under the parcels tray in the facia.

Our test car was supplied by the Ford Comany of Australia, Homebush, N.S.W.

"Wheels" WORK TEST and analysis of the

Ford Zephyr

Don Brodie, Ford Motor Company's garage foreman in charge of test preparation, poses the ute for our photographer.

THE N.S.W. countryside was in the throes of bush fires when we tested this newly designed Zephyr coupe utility. The vehicle, as delivered to us, was painted a bright fire engine red, and to this day we do not know whether the considerable attention we attracted was due to the appealing lines of the Zephyr or the mis-placed gratitude of fire-conscious citizens.

Nevertheless there is no denying the purely aesthetic attraction of this workaday vehicle. It is actually a curious blend of styling innovations, but this does not become apparent until you study the various aspects of the design with considerable detachment. The front, for example, has a huge full width grille of the type once popular in American cars and known as the "dollar grin." In direct contrast is the tailgate and rear end assembly which is delightfully free of ostentation.

Front grille apart, the Zephyr has uncluttered lines, giving an elegant and even snappy appearance. Fire engine red or no, the utility is certainly an eye-stopper.

This much we knew, even before we opened the driver's door. A quick glance in and around the cab showed that the Ford designers had sensibly retained as much of the Zephyr sedan design as possible. They have altered only the rear section of the bodywork, stiffened the two sub-chassis frames, added three additional leaves to each rear spring, and of course equipped it with stouter tyres. All mechanical parts — and even the gearing — remain the same as the sedan.

The result is a fast, fairly lively, economical 8 cwt. utility, which, when laden with anything like its full capacity, gives a more than adequate road performance — though it is less of a top gear vehicle than is its stablemate, the sedan.

Surprise . . .

With 5 cwt. of sand aboard, the test vehicle scaled 33¼ cwt. 52 per cent of the weight was placed over the rear axle, and as a consequence the resultant road holding and tenacity on corners — particularly on gravel roads — came as one of the motoring surprises of the year!

In short, the aggregate of an accurate and light steering system, an extremely capable suspension and sports car like road manners, enable the six cylinder Zephyr to give a most impressive account of itself.

When discussing a double purpose vehicle such as a utility, one must necessarily consider it as both a passenger car and a work horse.

As a pasenger carrier, we welcome the pretty but non-ostentatious cab, the adequate room for three adults, the fresh air cooling system and the extremely lively road performance.

As a working vehicle, the load capacity of only 8 cwt. comes as something of a disappointment when one considers the overall length of the Ford (14 ft. 8 ins), the 80 b.h.p. engine, and the size of the rear loading tray.

Under-rated . . .

To be fair of course, we must point out that the sub frames and suspension system are rugged enough to cope with far heavier loads, and, very likely, the Ford designers, well aware of the omnifarious purposes to which the Australian utility is put, have deliberately under-rated their new product.

Another point which must colour any description of a utility is that they have long since become the work horses of all trades. They have a permanent place in dense city traffic, an established spot in the rural land-

Utility

A most durably built truck with virtually no vices, excellent road manners, an acceptable thirst, and a handy turn of speed.

scape, and a very mobile home in suburban districts. Clearly then, a successful utility must be provided with the necessary equipment to give both faultless service in outback conditions and rugged indifference to the knocks and bumps of city life.

To this end, the Zephyr has sturdy rubber rear bumpers (of which more anon), blinking trafficators, excellent dust sealing sealing, and provision for such accessories as radio and heater. From a driver's point of view, its only deficiency is the absence of an adequate rear mirror for city driving.

Quietness . . .

Performance-wise we could find little to fault on the Zephyr, and much to praise. Not the least of its virtues is the mechanical quietness of *all* the mechanical parts, and its pedigree road manners. From a design point of view, two small things irked us. First we acknowledge and

Excellent accessibility, heavy-duty air filtration, rugged mechanical components make Zephyr a vehicle ideally suited to rough country conditions.

The Zephyr ute handles remarkably well, even when laden. We deliberately took this corner in a glorious four wheel slide; found that the ute could be corrected and controlled with a flick of the finger.

Loading deck is hardwood with steel skid rails; tailgate drops to give eight feet load length. Actual tailgate, of heavy gauge steel is one of the sturdiest we've encountered; can be mauled severely without suffering damage.

Seating position is comfortable, and car rides well. Controls are light, positive, and easy of operation. Speedo was found to be surprisingly accurate.

applaud the fitting of rubber bumpers beneath the tail-lights. But why — oh why — did they not *mould* the rubber in the contour it would eventually take?

The way things are, a straight strip of rubber has been forced into a somewhat unnatural position. The bumpers look trim — on a new vehicle. But we strongly suspect that before many hard knocks are taken the stresses in the rubber bumpers will cause them to distort and possibly to crack.

Our second grudge is more important, for it concerns road safety. No interior mirror is fitted to the vehicle, even though the rear window space is generous. Of course, a mirror is fastened to the right hand door. Even if this mirror was good, rear vision would not be the best — but as it is, the mirror on the test car was loose on its ball joint; and, since there is no means of tightening it, the mirror serves no purpose.

In our book, a utility should have two mirrors — one inside for travelling unladen, the other outside, and this we suggest would be a major improvement on the Zephyr.

Good Speedo . . .

Many small points, however, did meet with a rousing round of approval. The 36 foot turning circle is excellent for a vehicle of this size; the spare wheel cover is hinged, so that it cannot be left on one side when the wheel is hastily changed; the speedometer is large, extremely accurate, and the needle rock steady. What a pity, though, that it has no trip meter!

Forward vision is excellent, with both wings visible; and the anti-glare plastic covering on top of the facia panel will save many a pair of eyes from unnecessary strain.

For the purposes of our test, we loaded six bangs of sand, weighing a total of 5 cwt., on to the broad wood-and-metal tray. It seemed to us at the time that this load would be adequate to give the ute a thorough working test — yet it should not be too much to dampen overly the engine's natural exuberance. As things worked out, this proved to be correct.

The load tray is made from hardwood, and is lined with husky steel skid strips. The tray width is four feet exactly, and, with the tail gate closed, its length is 6 ft 2 ins. Open, the overall stowage space is extended to 8 feet — a very useful load carrier indeed! The side panels are almost 19 inches high, and the tail gate is four feet wide. A weatherproof is four feet wide. A weather proof tonneau cover comes as standard equipment.

Needless to say, the wheel arches eat into the load space; but no more than on any similar vehicle and indeed less than on some. The tailgate is supported by two sturdy hinges, with excellent rattleproof locking device, and robust, plastic covered support chains.

Elegant . . .

The passenger cab is identical with the sedan. Its facia is unusual in that it slopes forward underneath and although this styling innovation adds to its appearance, it does detract from the roominess of the glove box. Conscious perhaps that the glove box is unnecessarily small, the designers have added a large parcel shelf to the left of the underside of the dash panel, and somehow this manages to look rather out of place in its elegant surroundings.

Behind the bench seat is a generously dimensioned parcel shelf.

The cab is remarkably attractive, as we have stressed before. It has neat, washable plastic trim, a very tidy dashboard, and a commendably restrained quota of chromework. The bench seat — which is adjustible — offers ample room for three man sized adults, with leg and head room to match. Passenger comfort is unquestionable and — hurray — the doors are equipped with pulls you can really get hold of.

Pendant foot pedals are set sensibly, so that leg movement is kept to a minimum. The three instrument dials — a 0-100 m.p.h. speedometer, an ammeter, and a fuel guage — are placed directly in front of the driver.

Noisy Starter . . .

The test car started instantly on all occasions and in view of its mechanical quietness, we were rather surprised to hear the staccato whine of what must be the noisiest Ford starter since the Model A.

Mechanically speaking, the Zephyr rates top marks. Even the non-synchromesh first gear is quiet in use, and the six cylinder engine lopes through the hardest work with conspicious ease and reticence.

With its 5 cwt load, the Zephyr cruised happily at 60 m.p.h. It wheeled around tight bends with complete confidence, and showed no tendency at any time to wag its tail on bitumin.

On gravel, of course, the traction was less, and we found no difficulty in breaking the tail loose. This is, of course, to be expected with any vehicle, but it is to the credit of the Zephyr that it remained completely controllable at all times.

On more than one occasion, as the photo shows, we came round a corner in a full four-wheel slide, with no sense of insecurity at all. In fact we quickly became more confident of the Zephyr's handling qualities than we are of some cars deliberately designed for the sporting motorist!

After pounding the Zephyr over some atrocious roads, and taking many tarmac corners at speeds near the limit of adhesion; we can say emphatically that both road holding and suspension are far above average.

Strong . . .

The suspension is so strong and good that despite a particularly vigorous pounding over some really rutted roads, no shock crept up the steering column — nor did the suspension bottom.

In other departments, too, the car surprises. The steering is lighter and more accurate than we have experienced on a Ford before. It is, however, heavy when parking or manoeuvring. The gear change, though not outstandingly light, is certainly positive, and has first class synchromesh.

In the matter of gearing, the utility has, as we mentioned earlier, the same gear ratios as the sedan. This gives it the advantage of better fuel economy and a reasonably fast top speed. But the first ratio is definitely too high to start the laden utility comfortably and without strain uphill. Likewise, top gear is high enough to be used more as an overdrive than a normal top gear and since the box has but three ratios, the driver can accordingly find himself at a disadvantage on reasonably slight grades.

Lights are exceptionally powerful and have a wide flat spread. The dipper switch, to the left of the clutch pedal, is intelligently placed and acts positively.

The brakes are powerful and positive, requiring relatively light pressures. Fade is average but the brakes recuperate quite normally and when required will bring the vehicle to a brisk halt.

Summing Up . . .

SUMMING UP: The Zephyr is a likeable, lively, and lithesome vehicle; a modest but strong load carrier. It is a vehicle with very few faults and with a surprising number of virtues. Is keenly priced. ●

Rear of ute is commendably unadorned; tail and stop lamps are neatly grouped together in one housing. Rubber quarter bumpers would be effective but are questionably made (see text).

FORD IMPROVEMENTS

As this issue goes to the bookstalls, readers will have noted that two facelifted versions of Ford cars are beginning to make their appearance.

The Zephyr has acquired a new look in front, a revised grille with smaller "checkerboard" openings having replaced the previous grille depicted in this test. Some interior improvements also have been effected; notably the enclosure of the gearshift mechanism within the steering column housing and a rearrangement of the facia. An armrest has appeared in the middle of the front seat. Steering is now of the recirculating ball type.

The Prefect has been substantially revamped, with abbreviated tailfins, a chromed grille, and a new, larger rear window. Interior improvements are mainly confined to details. The facia is now enclosed in a binnacle before the driver and is a distinct improvement. ●

PERFORMANCE AT A GLANCE

TEST CAR:
Ford Zephyr Coupe Utility, supplied to "Wheels" by Ford of Australia. Price £1145 (including sales tax). Immediate delivery.

MAXIMUM SPEED:
Laden 77.2 m.p.h., best time, average of two runs 76.8 m.p.h. (Speedometer indicated 80 m.p.h. during fastest run.)

BEST ACCELERATION:
Standing quarter mile (laden) 23.6 secs; 0-30 m.p.h., 6.4 secs, 0-50 m.p.h., 16.3 secs.

MAXIMUM SPEEDS IN GEARS:
1st 33 m.p.h.; 2nd 54 m.p.h.; top 77 m.p.h. Recommended shift points 20 m.p.h. and 40 m.p.h.

BEST HILL CLIMBING:
1st gear: 1 in 4.8 at constant 25 m.p.h.; 2nd gear 1 in 8.0 at constant 35 m.p.h.; top: 1 in 13.5 at constant 30-40 m.p.h.

BEST PULLING RANGE:
Top gear, 30-40 m.p.h., peaking at 32 m.p.h.

BRAKES:
Footbrakes from 30 m.p.h., gear in neutral, stopping time 2.2 secs. (laden). Fade average with normal recuperation. Handbrake requires 7.2 secs to halt car from 30 m.p.h. in neutral.

ROAD BEHAVIOUR:

Type of Surface	*Behaviour*
Bitumin highway:	Excellent. Perceptible roll on tight corners.
Unsealed roads:	Above average.
Corrugations:	Faultless.
Pot holed winding roads:	Very little road shock up steering column.
Loose sand:	Good.
Mud, wet grass:	Good.
Cross country:	Excellent.
General riding comfort:	Excellent.
Dustproofing:	Excellent.

ECONOMY:
Hard driving test 20.3 m.p.g.; normal fast cruising 26.4 m.p.g. Car laden with 5 cwt. load in both tests.

TEST WEIGHT:
Fully laden, with 2 passengers, test gear and 5 cwt. of sand bags, 33¼ cwt. Distribution, 52 per cent rear, 48 per cent front.

CONDITIONS:
Fine weather, medium strength winds. All types of road surfaces encountered except deep water.

The ZODIAC in detail

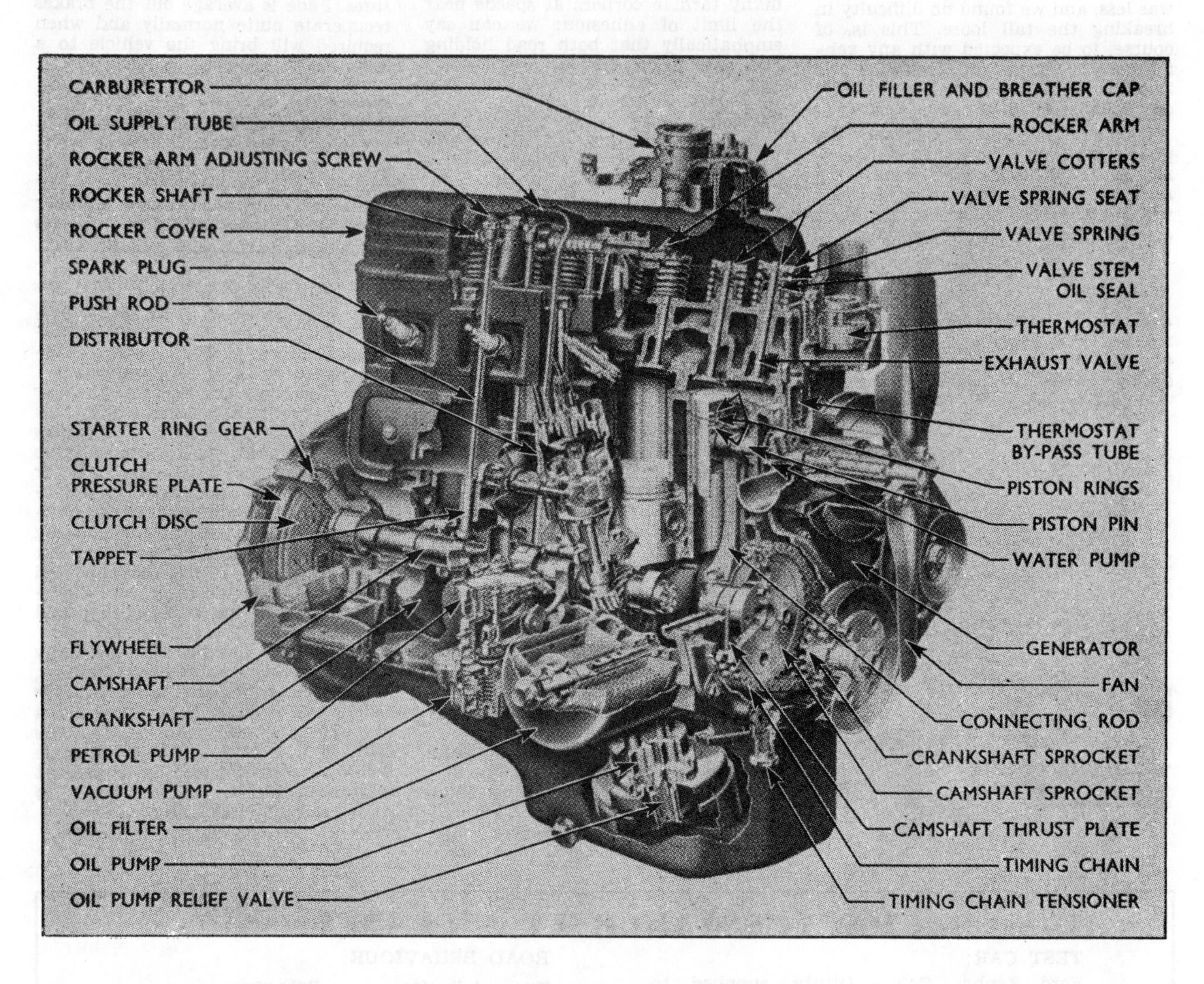

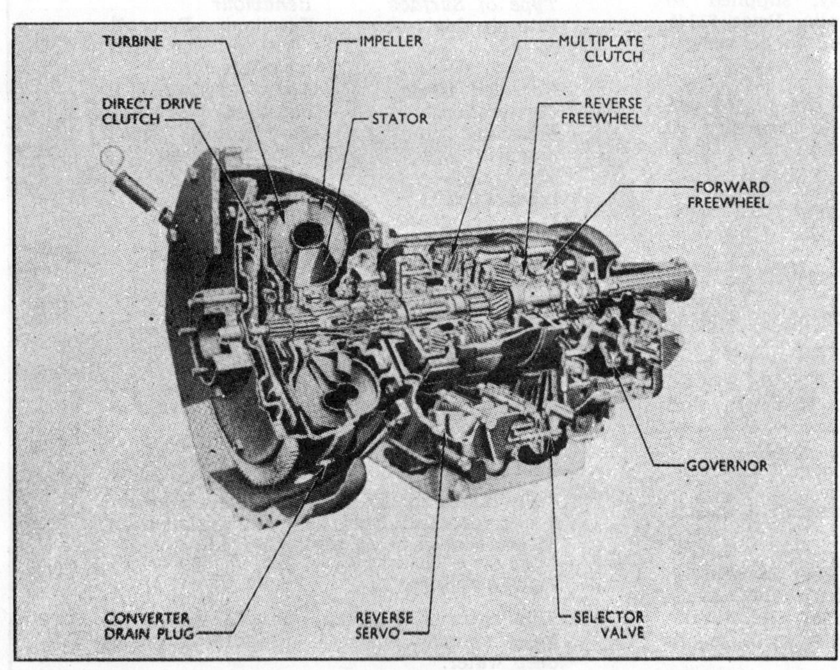

All the major components of the Zodiac's power unit are visible in this cut-away view (above). With a compression ratio of 7·8 to 1 this 6-cylinder 2,553 c.c. engine develops 90 b.h.p. (gross) at 4,400 r.p.m. Maximum torque is 137 lb.-ft. at 2,000 r.p.m., which is equivalent to 38 m.p.h. in top gear.

The Borg-Warner automatic transmission system (left) incorporates epicyclic gears and a torque converter. Top gear is direct; the ratio of the rear axle is 3·9 to 1.

A CAR ROAD TEST

The Ford Zodiac

with automatic transmission

THE Zodiac with Borg-Warner automatic transmission has a very wide appeal; it holds particularly strong attractions for the cost-conscious motorist who requires a roomy, good-looking, fast, quiet-running saloon car that is remarkably easy to drive.

Its limitations are few: it is, of course, not a sports car; it offers little to attract the motorist who is a keen driver in the sense of liking to exercise his judgment and skill in negotiating curves and bends with zest and in operating gear lever and clutch pedal adroitly. (A 'standard transmission' model is, of course, also available. So, too, are various means of increasing the power of the engine. This report deals with the Zodiac with standard engine and automatic transmission.)

Much of this Zodiac's character arises from the fact that its designers have sought to provide a generous sufficiency in all directions, and excess in none. Thus, for a coast price of £1,076, one has a six-seater car of elegant appearance that is extremely easy to drive and is capable of exceeding 85 m.p.h. in slightly favourable circumstances.

A motorist who invariably travels long distances with five passengers might elect to have a little more room; a motorist who seeks to cruise whenever possible at above 70 m.p.h. might seek yet more power. But the accommodation, power and speed of the Zodiac certainly cater more than generously for the requirements of the vast majority of motorists.

A carefully planned all-round balance of virtues is the aim of all who design vehicles to be made (and sold) in considerable quantities. In its particular price group, the Zodiac, in our opinion, excellently attains its target.

People with quite widely differing tastes may choose to buy this car for entirely separate reasons—and all be well pleased by their choice. Their reasons might be the Zodiac's lines, the seating accommodation and baggage-space that it offers, the ease with which it is driven, the widespread availability of specialised servicing and genuine spare parts that exists because it is a Ford product, its performance and reasonable petrol consumption—any one of these, or a combination of any two or more.

The Zodiac's 90 b.h.p. (gross) 2½-litre 6-cylinder engine mates very well with the Borg-Warner transmission system which provides automatic changes between the following three conditions: a direct-drive ("solid") top gear resulting in an overall ratio of 3.9 to 1, an intermediate gear in which the torque converter provides overall ratios ranging between 5.6 and 11.2 to 1, and a low gear in which the torque converter provides overall ratios ranging between 9.0 and 18.0 to 1. The gears are of epicyclic type and are quiet in operation.

The transmission lever may be set in any one of five lettered positions: P, N, D, L, R—Park (in which the

transmission is locked), Neutral, Drive (in which all normal motoring is done), Low (in which only the lowest range of ratios is operative), Reverse.

As a safety precaution, the wiring of the circuit of the starter motor is such that the engine may be started only when the transmission lever is set at either Neutral or Park.

Upward changes from the low gear to the intermediate, and thence to the direct top gear, are automatically effected according to road speed and throttle-opening.

Downward changes take place automatically when road speed falls—and, at all suitable higher speeds, they may be made by the driver depressing the accelerator pedal to the furthermost limit of its travel (this actuates an electrical switch which is positioned under the pedal).

In the full-throttle condition upward changes are delayed beyond the normal point; thus, with the accelerator pedal fully depressed, the change from low to intermediate occurs at 30 m.p.h., whereas this change is otherwise made at a point between 10 and 20 m.p.h. which is dictated by the precise degree of throttle-opening.

The change from intermediate to the direct top gear occurs at between 20 and 35 m.p.h. unless the accelerator pedal is fully depressed—in the latter ("kick-down") condition this change is made at 55 m.p.h.

These speeds and details are chiefly of academic interest. The motorist who elects to have automatic transmission will drive a thus-equipped Zodiac without thought of the ratio that is operative at any given time. With full justification, he will say to himself "Messrs. Ford and Borg-Warner know best", and, applying suitable pressure to the accelerator or to the brake pedal according to road conditions, he will obtain all the results he desires without conscious consideration of the ways and means involved.

The changes are always smoothly made, but particularly so at the lower road speeds. This is appropriate enough: when the car is being accelerated rapidly (when pressed it will reach 60 m.p.h. from a standstill in 18 seconds) the slight forward surge that accompanies upward changes is reassuring rather than disconcerting.

As is usual with automatic devices, the truly skilled, enthusiastic — and possibly rather 'old-fashioned' — driver may occasionally feel that, given a gear lever, he would have changed gear at a more suitable moment

Engine layout is neat and accessibility is good.

than the automatic system did. This type of motorist is not really in the market for automatic transmission, but even he will probably recognise the immense value of the system in town driving—and will admit that, within the limits imposed by the fact that no automatic transmission is capable of anticipating future events in the manner that an observant driver can, the system works well-nigh perfectly.

An automatic transmission system which provides two--pedal control contributes generously towards making a car easy to drive—but it does not, of itself, inevitably result in a vehicle that is driven without anxiety or strain. The view-out that is obtained from a car's driving seat, the positioning of its controls and the efficiency of its brakes also have great influence in this matter. The Zodiac scores well in these directions.

Visibility from the driving seat—and, indeed, from all seats—of the Zodiac is excellent. The driver has within his view the extremities and large portions of all four wings; he is able to execute parking and similar manoeuvres with exactitude and confidence. The brake pedal is sufficiently wide to encourage brake-application by means of the left foot when the car is being "inched" forwards or backwards.

The engine of the Zodiac that was submitted for test had a very quiet and utterly smooth idle, and there was no "creep", i.e. no tendency for the car to move forwards when the transmission lever was set in Drive position and the vehicle was standing still on a level road without pressure being applied to the accelerator pedal.

The Zodiac's brake pedal is set high in relation to the position occupied by the accelerator pedal; it is, in fact, so high that a tall driver may find the upper part of his right leg striking the rim of the steering wheel when he raises his foot to apply the brakes—until he acquires the habit of angling his knee to avoid this encounter.

With this exception, all of the Zodiac's controls are admirably positioned to ensure ease of operation by drivers of virtually all heights and build.

Completely free from any sign of fade in all normal conditions, the Zodiac's brakes provide very good stopping power in response to quite light pedal pressures. Their efficiency and consistency of performance contribute towards the driver's relaxed, peaceful state of mind. It is easy for the man or woman at the controls of this car to obey the Road Safety Organisation's injunction to drive relaxed and alert . . . and with a smile.

The seats are comfortable; little fatigue is experienced by driver or passengers after many hours of continuous motoring.

Moderately firm, the suspension system bestows almost roll-free cornering ability at normal touring-car speeds and a good ride, without any suggestion of pitching, over all types of surfaces. Bumpy roads can produce a certain amount of vertical motion at the rear (particularly if the baggage boot is empty), but the relative periodicities of the front and rear springs are well-planned to prevent this from developing into a pitching motion.

Starting is immediate. After a start from cold it is best to warm up the engine for a minute or two before driving off; this removes the possibility of stalling which otherwise exists.

Willing and smooth-running, the Zodiac's engine is most pleasantly unobtrusive at all times except when the car is being driven somewhat ruthlessly with the objective of extracting its maximum performance. A cruising speed of 65-70 m.p.h. proves agreeable on fast, national roads, and the Zodiac holds a straight course notably well even when quite a strong side-wind is blowing. At 70 m.p.h. crankshaft r.p.m. in top gear is 3,700.

The wide brake pedal may be operated by either foot; the pedal to the left of it conceals — and pressure on it operates — the dipper-switch of the headlights.

The baggage boot is deep and has an excellent capacity.

Standard equipment includes an electric clock, folding arm-rest to front seat, two-tone finish (monotone also available), and windscreen washers, cigarette lighter and whitewall tyres. Both front doors are fitted with key-locks. The test car was equipped with a radio, which is an optional extra. The vacuum-operated windscreen wipers have the virtue and the fault common to their type: they operate quietly, but they slow down almost to a halt when the throttle is opened abruptly.

With first-grade (90-octane by research rating) fuel in the tank, there is no suggestion of pinking at any time.

In the course of average motoring we obtained 24.3 m.p.g. Experiments conducted over shorter distances indicated that as much as 30 m.p.g. may be obtained by very gentle driving, and that only truly hectic driving will result in 20 m.p.g. or less.

Ably planned to provide precisely the type of motoring that very many modern motorists want, the Ford Zodiac is a versatile and likeable car which exemplifies the advantages of volume-production by offering excellent value for money. ●

SPECIFICATION AND PERFORMANCE

BRIEF SPECIFICATION

Make ... ... FORD

Model Zodiac with automatic transmission (Borg-Warner).

Style of Engine Straight 6-cylinder. Water-cooled. Overhead valves (push-rods).

Bore ... 3·25 ins. (82·55 mm.)

Stroke ... 3·13 ins. (79·55 mm.)

Cubic Capacity 155·8 cu. ins. (2,553 c.c.).

Maximum Horse-Power 90 b.h.p. (gross), 86 b.h.p. (net), at 4,400 r.p.m. (Compression ratio 7·8 to 1).

Brakes ... ... Hydraulic

Front Suspension Independent. Directly-operated coil springs mounted on double-acting dampers integral with wheel spindles. Anti-roll torsion bar.

Rear Suspension Semi-elliptic leaf springs.

Transmission System Borg-Warner automatic transmission, with manual control to select: normal drive range, low range, reverse, neutral, park.

Gear Ratios (overall) Low 17·96 to 8·981 to 1
Intermediate 11·19 to 5·596 to 1
Top 3·9 to 1 (direct)
Rev. 15·67 to 7·835 to 1

Final Drive Ratio ... 3·9 to 1

Overall Length ... 15 ft. 0½ in.

Overall Width ... 5 ft. 9 ins.

Overall Height ... 5 ft. 2 ins.

Ground Clearance ... ... 6¾ ins.

Turning Circle ... ... 36 ft.

Dry Weight ... ... 2,820 lbs.

Price ... £1,076 at Coast Ports

PERFORMANCE

Acceleration 0-30 m.p.h. 6·0 secs.
0-40 m.p.h. 9·2 secs.
0-50 m.p.h. 12·8 secs.
0-60 m.p.h. 18·0 secs.
0-70 m.p.h. 28·0 secs.
0-80 m.p.h. 42·1 secs.

With selector set at Drive, and using kick-down to obtain maximum acceleration:

From a steady 10 m.p.h. to 30 m.p.h. ... ... 4·6 secs.
From a steady 20 m.p.h. to 40 m.p.h. ... ... 6·0 secs.
From a steady 30 m.p.h. to 50 m.p.h. ... ... 7·0 secs.
From a steady 40 m.p.h. to 60 m.p.h. ... ... 8·2 secs.
From a steady 50 m.p.h. to 70 m.p.h. ... ... 15·0 secs.

In top gear from a steady 30 m.p.h. to 50 m.p.h. ... ... 9·5 secs.
In top gear from a steady 40 m.p.h. to 60 m.p.h. ... ... 11·0 secs.
In top gear from a steady 50 m.p.h. to 70 m.p.h. ... ... 15·8 secs.
In top gear from a steady 60 m.p.h. to 80 m.p.h. ... ... 23·5 secs.

Maximum Speed ... 83·6 m.p.h.

Maximum Speed in Intermediate Gear ... ... 55 m.p.h.

Fuel Consumption ... 24·3 m.p.g.

Test Conditions Sea level. Moderate wind. Dry road. 90-octane fuel.

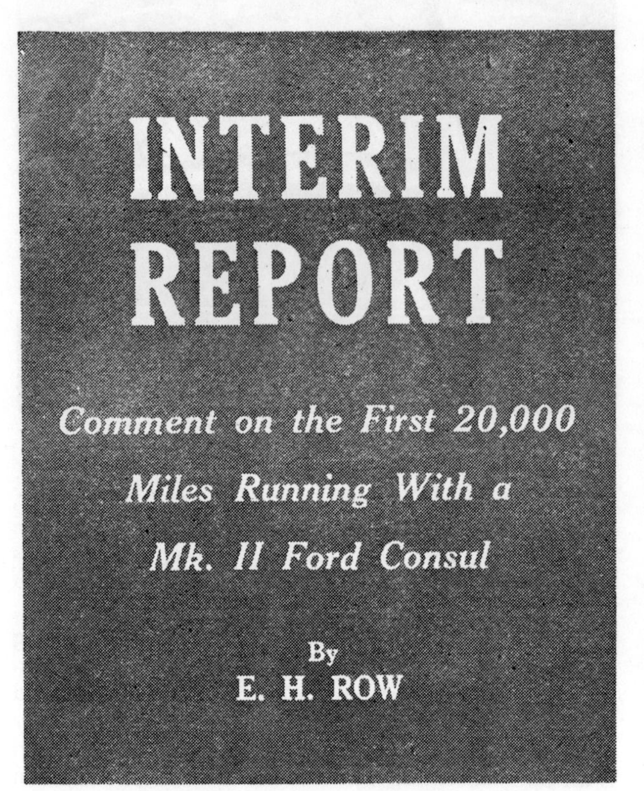

INTERIM REPORT

Comment on the First 20,000 Miles Running With a Mk. II Ford Consul

By

E. H. ROW

IT was a rule at one car manufacturing company, with which I served some time during the years between the wars, that a car being prepared for road testing by the technical Press, while it was to be standard in all mechanical respects, was nevertheless to be "just as good as standard could be made." As a result, several weeks were frequently spent in bringing the car up to a condition where it could be presented for unbiased criticism and produce the sort of comment that might increase sales; consequently, it was not unknown for quite a number of major component replacements to be made in the process.

So far as the average car of today is concerned, modern production methods have practically eliminated wide variations between one car and another from the same assembly line, with the result that cars now submitted to *The Motor* for Road Test, although probably "better brought up" than many which start from scratch in the hands of the average private owner, can generally be regarded as typical of the breed. But even my colleague Joseph Lowrey, with whom I sometimes find myself in argument, will scarcely disagree that, while a Road Test can give a cold appraisal of the general and detail performance of a carefully used car in its mechanical hey-day, it cannot aspire to predict how that car will behave over an extended period—say 20,000 miles.

So, as I have managed to keep a fairly detailed record of the running and general behaviour of the "straight-off-the-line" Mk. II Ford Consul which I acquired in March, 1958, and which, at the time of writing, has covered over 21,000 miles, such remarks as I have to make may be of interest, particularly if taken as complementary to those which have appeared in Road Tests of various examples of the marque published in *The Motor* from time to time: and particularly so when it is borne in mind that a car which serves a motoring journalist for everyday transport usually leads a harder life than most.

One's assessment of personal transport is generally governed by what has gone before, in this particular case a Triumph 2000 Roadster, nine years and 92,000 miles old when we parted company. Except for the fact that each has a four-cylinder engine and three-speed gearbox, there is little similarity between the two. HRW 60 had already done service in the hands of our Midland Editor before I took it over, so that I am unable to comment on the earlier stages and teething troubles, which had all been dealt with before it came down to London. It was, however, a car of character which, for most of the 60,000 miles that I ran it, served me well in a variety of capacities—for hurried, long-distance runs, as town transport, for country ambles with five up, and as a caravan tug, for which purpose its wide torque range made the Triumph eminently suitable. It was a thirsty beast—the best long-run average I ever obtained was 21.8 m.p.g. and overall about 18. Its Girling Hydrostatic brakes, while extremely effective with but the lightest of pedal pressures, needed reshoeing at approximately 7,000 miles and, by their constant contact design, undoubtedly contributed to the car's high fuel consumption. Nevertheless, with its tubular chassis, coachbuilt body, robust build and general comfort, coupled with a most acceptable touring performance, the Triumph was a car that engendered considerable affection. Its basic price in 1949 was £775; the addition of purchase tax brought the total up to £991. Ten years later the average second-hand price being asked for a 2000 Roadster is £358, or almost half its original basic figure—an indication of long-standing appeal. The main reasons for disposal were rapid deterioration of ancillary equipment, particularly electrical, and the fast growing cost of replacement parts. But the separation was a sad one.

Made for Work

The Mk. II Consul is a bird of very different feather; it has none of the Triumph's character nor, I think, will it engender quite the same degree of affection as its predecessor. On the other hand, I am sure that it is far less likely to cause those bursts of irritation that were wont to arise when the Triumph's petrol bills came to be paid, maintenance charges to be met or repairs to be done with increasing frequency. No, 909 PMK falls into the same category as the refrigerator, washing machine and vacuum cleaner—an unspectacular piece of machinery which does its job efficiently, comfortably and with

". . . a pair of Sparto reversing lights . . . fit tidily and invulnerably beneath the rear bumper . . ." These are seen above, as well as the Witter towing bracket, while on the right, a pair of Discus lamps fit well behind the leading edge of the front bumper where they are protected from major damage.

". . . conditions generally were worse and mileage covered greater than any previously . . . yet, at the end we were less car-tired than ever before."—Nearing the end of the 1958 Tulip Rally, the Consul's first major assignment.

reasonable economy; and it fits unobtrusively into the modern scene.

In the light of necessity and of previous experience, certain extras were included on the order—a centre folding armrest, not so much to rest the arm as to act as a roll-stopper for the front-seat passenger during fast-cruise cornering, a Smiths fresh-air heater and a screen washer. Subsequently, a pair of Butler's Discus lamps were added, a foglight which works very well indeed and a spotlight which is less good but balances the frontal aspect. Reason for this choice was the shallow dish of these lamps—2 in. maximum, enabling them to be fitted easily into the 4-in. space between the front of the bumper and the radiator-grille frame with the glass still 1½ in. behind the front bumper line. And because to reach my garage I have to reverse up a narrow 40-yard entry, a pair of Sparto reversing lights which fit tidily and invulnerably beneath the rear bumper, were also added.

Although the technical people at Dagenham take a poor view of their cars being used for towing caravans, the number of current Fords so employed suggests that customers, having motored solo through the guarantee period so as not to become "out of benefit," thereafter couldn't care less what the technicians like or don't like. So, as one's job entails occasional caravan towing, the Consul was equipped with a Witter towing bracket and a pair of Widavue extending wing mirrors.

Finally, the underside, the insides of the doors and the luggage boot were given a heavy coating of undersealing compound, both as a protective measure and to reduce drumming which otherwise occurs on certain types of road surface and at certain speeds. As a result, the car is quite quiet under most conditions, resonance only occurring on the rougher-cast types of top dressing between 50 and 65 m.p.h. This is with 5.90-13 Goodyear tubed tyres, inflated to 26 p.s.i. back and front. They have just been removed, still with signs of tread remaining, and sent for retreading; the replacements are of the same make.

Ordinary Treatment

From the outset, the Consul being a commonplace car (over 200,000 have been sold at the time of writing) has been treated in a commonplace manner, receiving the prescribed servicing at the prescribed intervals provided it was convenient, but not coddled or fiddled with.

Running-in having been carried out in accordance with maker's suggestions, the first real job of work involved coverage of a Continental rally, a matter of some 1,500 miles in three days and a night. I had undertaken this task in the four previous years, using as transport a drophead Mk. I Zephyr, two open sports cars and a quantity-produced family saloon of slightly over 2½ litres equipped with Borg-Warner overdrive. In the 1958 event conditions generally were worse and mileage covered greater than any previously, yet both the co-driver/photographer, who has done this trip with me on most occasions, and I were agreed at the end that we were less car-tired than ever before. Subsequent long journeys have only served to strengthen my belief that in such conditions the Mk. II Consul takes less out of one than almost any other car in the popular price group. Both front- and back-seat passengers mostly echo the same sentiment.

At the time of writing, the mileage recorder on the Consul shows 21,307 indicated which, corrected for 2½% flatter, is an actual figure of 20,775 miles, of which, according to my records, 11,745 have consisted of long-distance journeys and the remaining 9,030 of local town and suburban use. It was

INTERIM REPORT

The good, but not obtrusive lines of the Consul are emphasized by early morning light in this photograph of the car in the Jura Mountains.

my original intention to compile this report after an initial year's running and I did, in fact, work out the overall fuel consumption for the 14,508 miles covered in the period. These, broken down into long-distance and local-running figures, totalled 7,450 and 7,058 respectively, and to do this 537 gallons of premium-grade fuel costing £126 8s. 4d. were required—an average of 25.1 m.p.g. or, in terms of cash, 2.1d. per mile. Unfortunately, records over the remaining mileage are incomplete. However, a recent 738-mile check, involving main-road and local town and suburban running plus the performance tests, showed an average of 22.4 m.p.g.

Extra Weight

The addition of the various bits and pieces, plus the under-sealing compound, has increased the weight of the Consul by 1¾ cwt. to almost exactly the same total as the 1958 convertible of which a Road Test was published in *The Motor* of January 8, 1958. The figures obtained on that occasion were, therefore, used as a basis for comparison with some obtained during a recent test of my saloon after 20,000 user miles. During that period the engine and other major components have remained untouched except that, immediately prior to the tests, the oil filter was changed (it was due to be anyway), the plugs were cleaned and reset and the distributor points properly gapped. The tests themselves were carried out over the same stretch of road as that used for the convertible but, whereas the temperature on the earlier occasion was around 45, the barometer stood at 29.1 and the weather generally was cool and damp, that obtaining when the saloon made its runs was hot and dry with a temperature of 75 and the barometer standing at 30. The wind was light on both occasions.

The comparative figures, which are shown in an accompanying table, reveal a notable falling-off at the top end of the speed range on the Maximile speeds and a corresponding deficiency in the top-gear acceleration from 50-70 m.p.h.; this suggests that the engine would benefit from a top overhaul, a third set of plugs and an ignition check which, indeed, it will shortly get. This presumption is further borne out by an average 5 m.p.g. drop in the steady-speed fuel consumption between 30 and 60 m.p.h., as well as by the overall increase. Excessive carbon deposit is unlikely to have occurred, the oil used averaging well over 2,000 m.p.g., which figure includes eight drainings and refillings of the 6-pint sump.

The hard driving which 909 PMK has undergone in covering two International rallies and on a variety of rather hurried journeys in the course of business has subjected the car in general, and the transmission in particular, to considerably greater strains than would be the lot of more normally driven cars, particularly as regards using the three-speed gearbox in a sports-car (which the Consul is not) manner for cornering. As a result, the clutch exhibited symptoms of unhappiness at about 14,000 miles, the cure being a new plate and fingers, and the rear universal joint on the propeller shaft needed replacing at just under 20,000 miles.

Probably because of the acknowledged tail-lightness of earlier Mk. I Consuls, which occasionally gave rise to something of a *crise* when roads were wet, one is not infrequently questioned regarding the stability of the Mk. II. In this regard, I have nothing but praise for 909 PMK. I have driven it many miles on snow, ice and grease without discovering any particularly nasty habits, and on dry roads it possesses the sort of road-holding and stability that makes a 65 m.p.h. cruising gait both safe and comfortable. Brakes are powerful under light pedal pressure and progressive in action, except for the first few applications after the car has been left standing for a while when, like those on several other current models, they tend to peck on the first few applications. Some attribute this to an accumulation of dust from the shoes, others to the formation of a slight film of rust on the drums, and yet others to the particular material from which the drums are made. Whatever the reason, a remedy is to drive for the first

FORD CONSUL—COMPARATIVE TESTS

	1958 Ford Consul Convertible Road Test (Weight as tested—27½ cwt.)		1958 Ford Consul Saloon after 21,000 miles (Weight as tested — 27 cwt.)	
"Maximile Speed"—timed quarter mile after one mile accelerating from rest				
	m.p.h.		m.p.h.	
Mean	77.8		73.6	
Best	79.6		75.0	
Acceleration Times—upper ratios				
	Top gear	2nd gear	Top gear	2nd gear
m.p.h.	sec.	sec.	sec.	sec.
10—30	12.1	6.7	12.6	7.4
20—40	11.8	7.7	12.0	7.4
30—50	12.7	9.2	12.9	9.8
40—60	15.6	—	16.2	—
50—70	23.0	—	27.0	—
Fuel Consumption				
	m.p.h.	m.p.g.	m.p.g.	
At constant	30	38	32.5	
At constant	40	36	29.6	
At constant	50	31.5	26.7	
At constant	60	27.0	23.7	
Hill Climbing—at sustained steady speeds				
Max. gradient on top gear	1 in 12.4 (Tapley 180 lb/ton)		1 in 11.7 (Tapley 190 lb/ton)	
Max. gradient on 2nd gear	1 in 7 (Tapley 315 lb/ton)		1 in 7.1 (Tapley 310 lb/ton)	
Instruments				
Speedometer at 30 m.p.h.	2% fast		1.7% fast	
Speedometer at 60 m.p.h.	3% fast		3.3% fast	
Distance Recorder	4½% fast		2½% fast	

"Generally, . . . the seats and trim show few signs of getting shabby, and the seat springs of losing overmuch temper."

INTERIM REPORT

50-100 yards with the *left* foot applying light pressure on the brake pedal.

Prior to departure for this year's Scottish Rally, the brake drums were removed for a precautionary examination of the linings. This was at 15,260 miles. There was probably another 5,000 miles of wear left in the shoes, distributed fairly evenly over all four sets. However, removal of the brake drums on the Consul entails withdrawal of the hubs with which they are integral so, in the interests of labour-saving, replacement shoes were fitted while the drums were off.

Reliable Dampers

Dampers generally are a much-maligned commodity. In some cases criticism may not have been altogether unjustified, but so far as my particular Ford is concerned it does not apply. Despite a hard life, these components (original fitting) are still standing up well to their job and have needed no attention other than topping-up at the prescribed intervals—perhaps that is the reason why.

One other point occurs to me on the mechanical side—the gear change. When the car was new there was occasional difficulty in going from first to second, and vice versa. The distributors, with whom the matter was raised, professed themselves unable to do anything about it, so there the matter rested and the difficulty has subsequently disappeared. That it has done so is obviously more due to increasing familiarity with the car than any process of elimination through running-in, for comment on the subject from colleagues who occasionally borrow the Consul is such as could not well be printed on these pages. Particular interest attaches to this criticism when it is remembered that, in the "face-lifted" Consuls that were introduced in February this year, the offending concentric type of gear-shift mechanism was superseded by a less temperamental type of change.

Nothing, so far, has been said about the bodywork and upholstery, matters more dear than mechanical efficiency to some classes of car owner. When not on the road 909 PMK spends much of the time standing in an open car park in a part of London where grime and fumes predominate and where, during this magnificent 1959 summer, there was no shade whatsoever. Compared with newer Consuls of similar colour, the Newark grey paintwork has changed remarkably little. Certainly there are odd scars on various parts of the body, the penalty of metropolitan motoring, but so far, except on the small uprights that separate the rear-door windows and the fixed quarter lights, "maker's rust" is notably absent. The chrome, too, has retained its brightness very well.

Whoever chose red and cream as the colours for the upholstery failed singularly to appreciate how these two shades show the dirt and how the matt texture of the P.V.C. clings lovingly to every spot of London grime. Repeated scrubbings with ordinary soap and warm water, and subsequently with well-known liquid detergents, only half-succeeded in removing the dirt; it was not until recently, when I discovered Decosol which you merely paint on to the upholstery, leave for five minutes and then wipe off with a slightly damp sponge, that the P.V.C. was restored to its pristine freshness. Generally, though, the seats and trim show few signs of getting shabby, and the seat springs of losing overmuch temper.

That, then, is my interim report on an everyday car after sixteen months of, frequently, more than everyday motoring. Up to now there appears no reason why a further report when the mileage has been doubled should be any less relatively favourable.

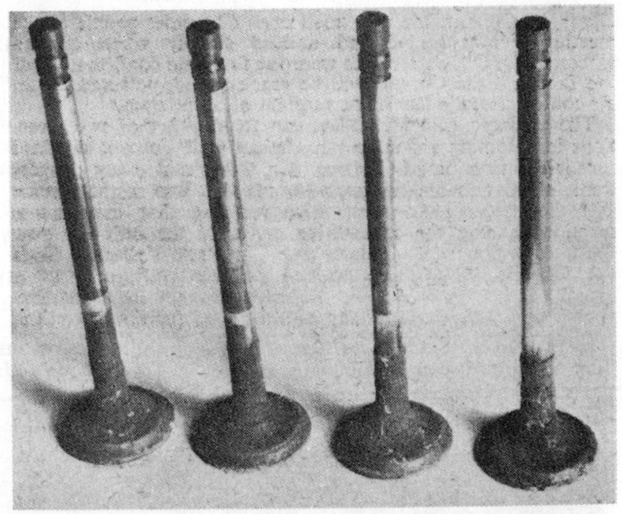

Where the power went: close-ups of the exhaust valves and one of the exhaust valve seatings, taken during the initial "decoke" after this article was written, suggest that here was a major reason for the power drop evidenced by the running tests.

2,000 Miles in Zephyr Wagon

ANYONE who tries to break into the station-wagon market in Australia has a hard task ahead of him.

Holden's station sedan dominates the field with an estimated 80 percent of sales — and Standard's Vanguard estate car is equally firmly entrenched as the No. 2 favorite.

But, after covering 2000 miles in an early-production model of the Zephyr station wagon, I am convinced that Ford's newcomer will make the grade.

Priced at £1425, including tax, it falls roughly midway between the Holden Special station sedan (£1259) and the Vanguard estate car (£1549) and should be able to carve off some sales from both rival territories.

The Holden wagon, to my mind, has an edge on the other two in looks and resale value, while the Vanguard has earned an enviable reputation for ruggedness and durability; but Ford's newcomer wins out on horsepower and all-round roominess.

To give a completely fair comparison of prices, I must mention that the Vanguard comes complete with air-conditioning and screen-washers, while on the other two cars these cost extra.

Consequently, to match Vanguard's equipment, you must add about £50 to the price of each of its rivals.

The Zephyr wagon I tested had all the trimmings — A.W.A. transistor radio, a superb Smiths air-conditioning unit, and my favorite safety device, screen-washers. These items added nearly £100 to its basic price, but made the wagon complete and as luxurious as anyone could possibly wish.

Getting Acquainted

When I collected the wagon from Ford's N.S.W. sales manager, Max Gransden, I knew I was going to enjoy this trip test.

The vehicle was definitely good-looking — white with red top and big, fat 6.70 by 13in. tyres giving a comfortable air to the silhouette. The outline was also quite rakish with the cruiser-like rear.

Inside, the dominant color was red, with touches of white giving a clean and spacious effect.

The wagon's front compartment is pure Zephyr — or rather Zodiac, as the new dash layout has been copied from England's luxury Ford. Wide bench seats front and rear will take six persons without any crowding — a rare thing these days.

There's generous space in the back for a family's luggage — but what a pity the spare wheel is buried under it!

The tailboard shuts in the normal manner, and then you wind up the rear window, locking it in any position you wish for ventilation purposes. This makes for quicker and easier handling of suitcases, and I can assure the sceptics that the window doesn't rattle.

Cockpit, Equipment

Behind the wheel, I found the repositioning of the starter-key-switch a big improvement. It is now on the right of the wheel, away from the choke; in its previous position, on the left side, you had to reach either through or around the wheel to get at it, and dodge the gear lever at the same time.

But why do Ford continue to fit those dreadful vacuum-type screen-

***DOORS** open wide on neatly upholstered interior, Zodiac-style dash layout. But why no armrest on driver's door?*

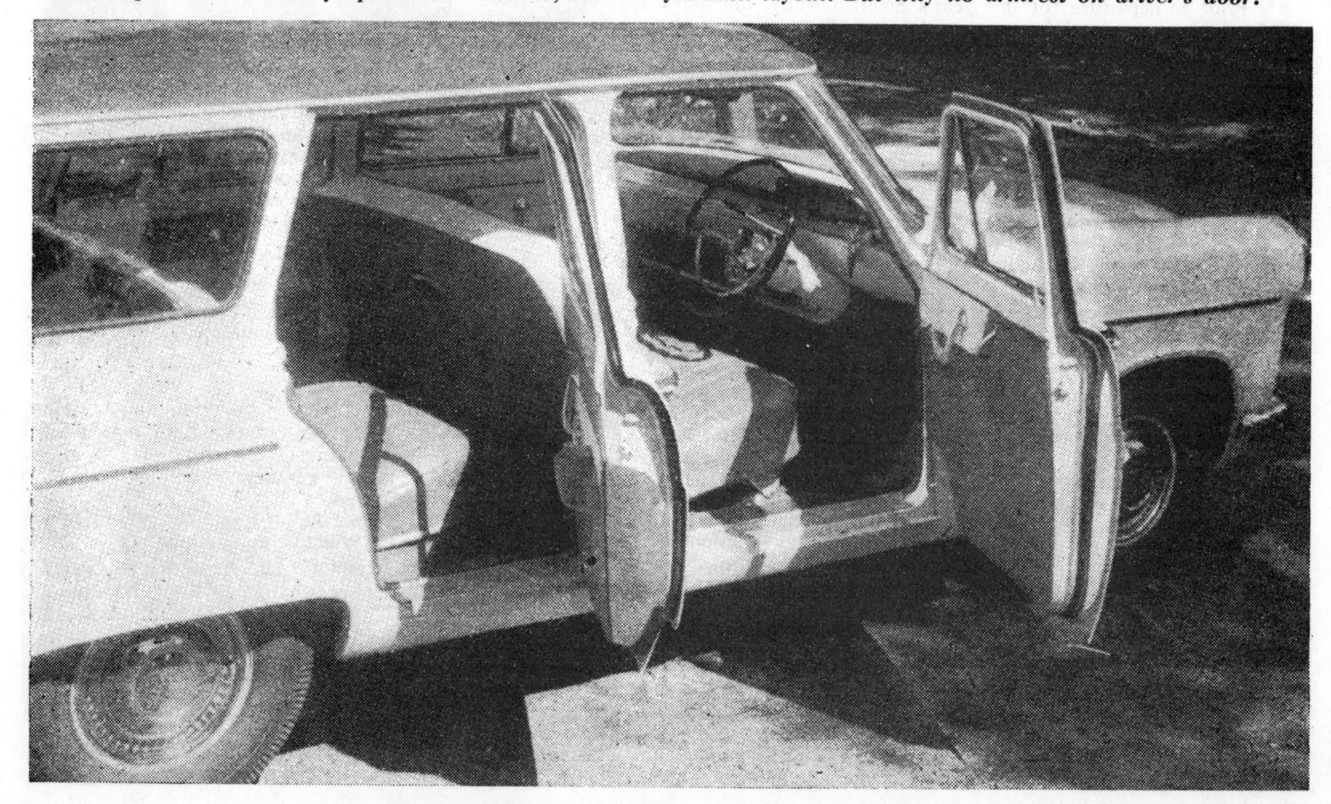

Fast, roomy and extremely comfortable, Ford's newcomer should garner a fair share of our tight station-wagon market, says tester David McKay

FOLDING the back seat was quite a chore; but car was an early model and system has since been improved.

wipers, complete with the insensitive control knob? I was to curse them frequently during the next fortnight! In the interests of safety, it is high time electric wipers were fitted.

The central armrest was much appreciated by all three of us. It made for comfortable long-distance driving, and my baby daughter could sit on it and see what was going on, giving my wife a break from holding her.

Strangely enough, the wagon hasn't got the door armrest fitted to the saloon. Instead there is a door-pull handle which can whack your funny-bone and is certainly a poor substitute for the armrest!

Women will appreciate the big parcels tray under the dash. The glovebox, too, is quite adequate in size — but its lining is furry in texture and rubs off on everything you store there. The instruments are neatly arranged in a cowling which

MAIN SPECIFICATIONS

ENGINE: 6-cylinder, o.h.v.; bore 82.5, stroke 79.5mm., capacity 2553 c.c.; compression ratio 7.8 to 1; maximum b.h.p. 86 at 4200 r.p.m.; maximum torque 136ft./lb. at 2000 r.p.m.; Zenith downdraught carburettor, mechanical fuel pump; 12v. ignition.
TRANSMISSION: Single dry-plate clutch; 3-speed gearbox synchromeshed on top two; overall gear ratios—1st 11.08, 2nd 6.40, top 3.90 to 1; reverse 15.06 to 1.
SUSPENSION: Front independent, by coil springs and enclosed hydraulic telescopic shock-absorbers; semi-elliptics with lever-type hydraulic shockers at rear.
STEERING: Recirculating-ball type; 3½ turns lock-to-lock, 37ft. turning circle.
BRAKES: Hydraulic, 2 l.s. front; lining area 147 sq. in.
WHEELS: Pressed-steel discs with 6.70 by 13in. tubeless tyres.
CONSTRUCTION: Unitary.
DIMENSIONS: Wheelbase 8ft. 11in.; length 14ft. 10½in., width 5ft. 9in., height 5ft. 2in.; clearance 6.8in.
KERB WEIGHT: 27cwt.
FUEL TANK: 10½ gallons.
DIFFERENCES between Zephyr station wagon and sedan (sedan in brackets); Tyres 6.70 by 13in. (6.40 by 13); rear springs, 7-leaf (6); height 5ft. 2in. (5ft. 1in.); weight (full tank) 27cwt. (25).

PERFORMANCE ON TEST

CONDITIONS: Fine, cool, no wind; dry bitumen; two occupants; premium fuel.
BEST SPEED: 90 m.p.h.
FLYING quarter-mile: 88 m.p.h.
STANDING quarter-mile: 21.0s.
MAXIMUM in indirect gears: 1st, 32 m.p.h.; 2nd, 65 m.p.h.
ACCELERATION from rest through gears: 0-30, 4.2s.; 0-40, 7.8s.; 0-50, 11.6s.; 0-60, 16.8s.; 0-70, 23.0s.
ACCELERATION in top: 20-40, 8.4s.; 30-50, 9.0s.; 40-60, 12.0s.
ACCELERATION in second: 20-40, 5.2s.; 30-50, 6.0s.; 40-60, 9.0s.
MOUNTAIN CIRCUIT: 51 m.p.h. average.
TEST HILL CLIMB: Two min. 35sec.
BRAKING: Before fade test: 88 percent efficient, 34.3ft. to stop from 30 m.p.h.; after fade test: 76 percent, 39.7ft. to stop from 30.
FUEL CONSUMPTION: 23.5 m.p.g. overall for hard-driven 2000-mile test.

PRICE: £1425 including tax

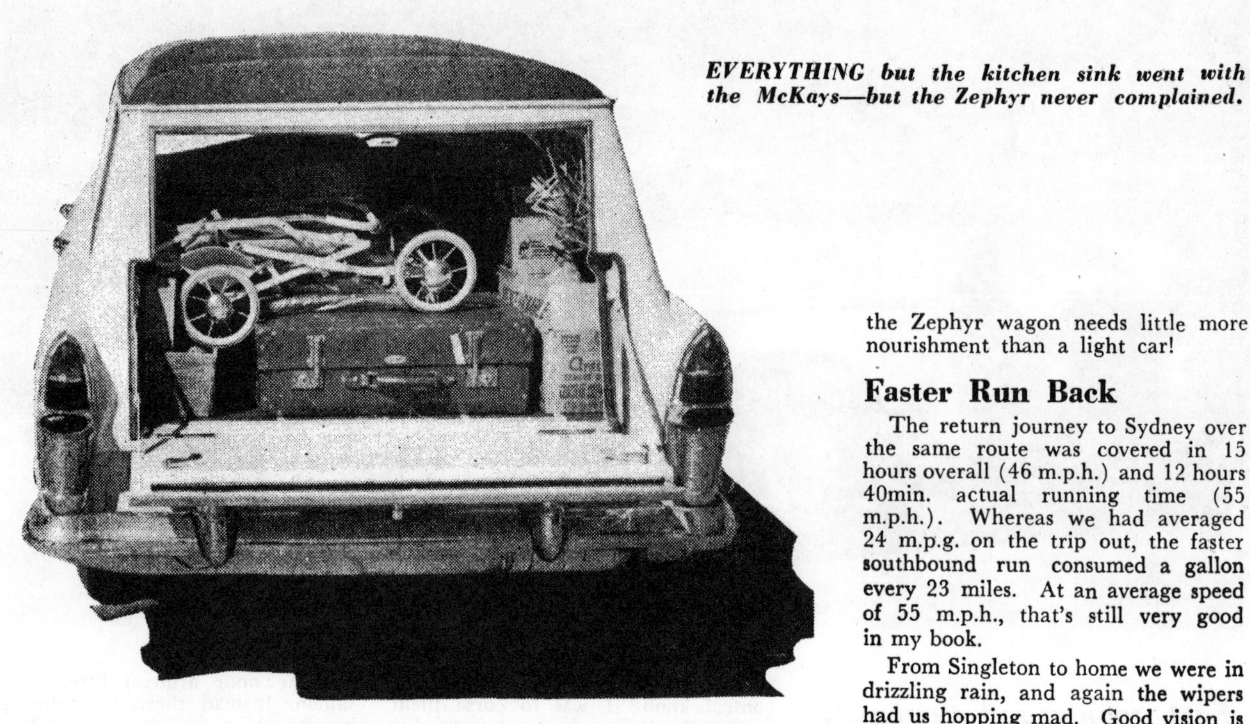

EVERYTHING but the kitchen sink went with the McKays—but the Zephyr never complained.

forms part of the full-width crash pad.

Getting the Zephyr home, I set about packing it with a playpen, cot, bath, high chair, several suitcases, cartons of tinned baby food, Dunlopillo mattresses, toys, thermoses, and a couple of dozen beer.

We were going to Surfers' Paradise for a week, and when I saw the quantities to be packed, I was glad of the wagon's vast interior. That is, once I got the seats folded properly. It wasn't an easy job to fold or erect the rear seats; but, as already mentioned, this was an early-production model, and Ford tell me the folding system has already been improved.

Fast but Thrifty

As we were planning to reach Surfers' in one day—700 miles from door to door by the speedo — we got cracking at 4.30 a.m.

Anxious to keep the baby asleep and to get the feel of the fully laden wagon, I took the Sydney-Newcastle section fairly leisurely — yet the Zephyr had us in Singleton by 7.30. Refuelling here showed our consumption to be 28 m.p.g. — another pleasant surprise.

Motoring more rapidly after Singleton to make up for a 30-minute stop, I found the engine was suffering from fuel starvation. At Muswellbrook we drove to the local Ford agents, who looked the Zephyr over and found a kinked plastic connection in the fuel line, out of sight beneath the battery.

We lost 45 minutes, but still made Tamworth by 10.45. Fuel mileage on this leg fell to 21.4. Leaving Tamworth at 11.15, we fuelled next at Glen Innes, and m.p.g. was as low as it was ever to get — 21. The fast run over the Moonbi Ranges — a delay for roadworks and the subsequent speed-up to maintain our schedule — made 21 m.p.g. quite a respectable figure.

After lunching at Armidale (12.45 to 1.30 p.m.) we fuelled at Warwick (Queensland), and the mileage rose to 24. Considering we had been cruising at over 70 m.p.h. most of the way, it was very satisfactory.

In Queensland

We left Warwick at 5.30 and reached Surfers' three hours later. Again m.p.g. was 24!

Shortly after Warwick we ran into a storm, and had a greasy road all the way via Cunningham's Gap to the Gold Coast. Despite the load, the Zephyr handled the nasty conditions most creditably, pointing where desired and never giving us a moment's worry as we hurried along.

But those vacuum wipers soon had me rubbing my eyes and peering into the darkness. Up any gradient, the damned things would stop working and leave me at the mercy of the rain-swept screen and the dazzle of approaching headlights.

Our north-bound overall time was 16 hours or 43.7 m.p.h. Running time was 13 hours, so our average road speed worked out at 53.8 m.p.h.

At Surfers' the Zephyr was given an oil-change and lube, plus a good wash. and drew its share of admiring glances in an area long used to flashy motor-cars.

While in Queensland, we drove up to see the Gold Star meeting at Lowood, and the wagon proved excellent for picnicking and viewing the race. We used the tailboard both as a table and a grandstand.

Cruising back from Lowood to the Gold Coast, we were amazed to find our fuel economy shoot up to 30 m.p.g. At speeds below 60 m.p.h., the Zephyr wagon needs little more nourishment than a light car!

Faster Run Back

The return journey to Sydney over the same route was covered in 15 hours overall (46 m.p.h.) and 12 hours 40min. actual running time (55 m.p.h.). Whereas we had averaged 24 m.p.g. on the trip out, the faster southbound run consumed a gallon every 23 miles. At an average speed of 55 m.p.h., that's still very good in my book.

From Singleton to home we were in drizzling rain, and again the wipers had us hopping mad. Good vision is essential at any time, but on the Gosford-Sydney stretch it becomes particularly important.

Total running costs for the 2000 miles were £19/14/4—and that includes the oil-change, lube and wash we got at Surfers'. Working out at less than £1 per 100 miles, this represents excellent value for your money in terms of fast, comfortable motoring.

Apart from criticism of the wipers, I have only one other complaint.

With its new 7.8 to 1 compression engine developing 86 b.h.p., the Zephyr is a fast machine and needs better brakes. Most people may consider the present stoppers adequate —but those who want to make full use of the Zephyr's performance will agree with me.

Lining area is large enough at 147 sq. in., but the drums are tucked away inside those 13in. wheels, where they don't get enough cooling airflow. Any Zephyr driven in competition is only limited by its brakes and nothing else. In engine, steering, suspension and general roadability I class the Zephyr ahead of its opponents. So let's have some stoppers to match the performance!

Put over my regular test route after the return from Queensland, the wagon recorded sports-car times for the mountain circuit, averaging 51 m.p.h., and shot up my "private" test hill in 2min. 35sec.

These are quite amazing figures—yet the Zephyr performed almost lazily, and I had the feeling it would have liked a harder workout.

Did I say it would make the grade on the station-wagon market?—I'll go a step further and predict that Ford's main worry will be how to meet the demand. ● ● ●

Used Cars on the Road—135

Three styling stars have been added on each side of car above the rear rubbing strip and there is a pointed bonnet mascot. Marks of a slight graze on the left front wing are the only blemishes on the paintwork, as most of the inevitable chips and scatches have been retouched

1956 FORD ZEPHYR II

Basic price new £580 0 0 *Total price new* £871 7 0

Price secondhand £695 0 0

Acceleration from rest through gears:

to 30 m.p.h.	5.2 *sec*	20-40 *m.p.h. (top gear)*	9.5 *sec*
to 50 m.p.h.	13.7 *sec*	30-50 *m.p.h. (top gear)*	10.3 *sec*
to 60 m.p.h.	20.6 *sec*		
to 70 m.p.h.	28.3 *sec*	*Standing quarter mile*	21.3 *sec*
Petrol consumption	24-28 *m.p.g.*	*Date first registered*	*November* 1956
Oil consumption	*negligible*	*Mileometer reading*	49,094

Provided for test by The County Garage, Gerrards Cross, Buckinghamshire. Telephone: Gerrards Cross 2279-3725.

VETTING used cars is usually simple and straightforward when vehicles of only two or three years of age are being considered. Within limits, the chief points to check are that the car has not been involved in an accident (in which case details of the repair must be known) and to ensure that the mileage covered is not excessive. This Ford Zephyr II is an example of a young, but high-mileage used car, and it is interesting that its condition reveals little evidence of the 50,000 miles travelled.

Occasionally a slight whiff of engine fumes is smelt inside the car, and under the bonnet the blow-by may be seen escaping from the oil filler cap. It might be worth while for the next owner to have a rocker-box breather fitted, because there are no other signs of wear in the car's mechanism. The engine starts well and warms up quickly from cold, though stalling occurs until normal temperature is reached. At all times the unit remains commendably quiet, the only cause for criticism being a degree of drumming at low revs. Throughout the speed range the engine pulls well, and the acceleration figures recorded are little lower than those of the model when new.

Some lack of precision has developed in the throttle linkage, and the result is that the clutch is inevitably subjected to rather fierce treatment; but there is no clutch-spin or judder. The gear box is quiet in all three ratios, and the synchromesh does its work efficiently. Second gear is occasionally difficult to engage.

There is good response to light pressure on the foot brake, which has adequate power for the not inconsiderable performance of the car. The approaching need for adjustment is indicated by the length of pedal travel. The umbrella-type handbrake is feeble and cannot be trusted to hold the car.

At about 50 m.p.h. a certain amount of front end shake is noticeable, felt as a tremor at the steering wheel, but this is a fault which may be cured by wheel balancing. The steering itself has developed a mild degree of free play, but the control is still precise at high speed, and the directional stability of the Zephyr is good. Although noticeably weaker than those of the model when new, the dampers still work efficiently: vertical pitching is limited, and the car gives an extremely comfortable ride. It corners reassuringly well, though there is some tyre squeal and rather more body roll than is usually associated with the Zephyr II.

Inside the car the slightly soiled and sagging appearance of the driving seat is the only visible confirmation of the mileometer reading. The left side of the front bench seat shows, to a lesser degree, similar signs of wear, but the excellent condition of the rear compartment suggests that back seat passengers have been rare. The upholstery is in white and maroon leathercloth. In both compartments the black carpets are clean and little worn, and the plastic roof linings are unmarked. Most of the painted interior trim is unblemished.

The finish is the creamy-white colour which is common on current Fords, and the first impression is of immaculate bodywork. Only close inspection reveals that in many places the cellulose has been neatly retouched. Chromium free from rust or scratches completes the picture of a well-kept, though admittedly young, used car.

All the tyres are Firestones, those on the front wheels and the spare being approximately half worn. The rear tyres have virtually unused Town and Country covers. There is a handbook in the glove locker, but the toolkit is strictly "jack only."

This Zephyr is most generously equipped with well-chosen accessories, which indicate that the one owner named in the registration book is a discriminating motorist. Borg Warner overdrive is fitted, with the usual accelerator-controlled kick-down switch to disengage the overdrive at full throttle. To supplement this, a facia-mounted switch has been wired in parallel, which enables the driver to cut out the overdrive without using full throttle. This arrangement proved to be very convenient—particularly when the car was fully laden—on the many occasions when speed fell too low for overdrive top gear, but full throttle was not appropriate. The fact that this is the first car tested in this series to be fitted with an overdrive is a reminder that this is a fairly recent accessory, not becoming really common until after 1954.

The standard fresh-air heater (of which the demister is not working) and radio are among the other accessories. There are also a cigarette lighter, Trico windscreen washer, two extremely powerful Marchal spot lamps, two wing mirrors and, mounted above the steering column, an oil pressure gauge and a coolant thermometer. This is still not the limit of the improvements on the standard model, as the underneath has been sprayed with a rubberized sealing compound, and lights have been fitted under the bonnet and in the boot. The under-bonnet lamp is of low power and connected to the side lamps switch. The boot light has its own switch, but is not working —the only part of the car's electrical equipment which is out of action.

It will not be easy for a prospective buyer to decide just how sound a purchase this car would be. The effect of the mileage on all mechanical parts is not to be underestimated, and a proportionate rate of general wear is the inevitable inference. On the other hand there are many indications which suggest that the car has had a very respectful ownership, and it is certain that it has been well maintained. The complete absence of squeaks or rattles, and the tautness of the car and its controls are noticeable at once; the mechanical condition is certainly much above the average found in cars which have run up this sort of mileage figure over the more usual period of about five years.

Condition, not history, is the important factor in this case. If allowance is made for the value of more than £100-worth of extra equipment, this Zephyr can only be regarded as very fair value at the price asked.

A thermometer and oil pressure gauge are the extra instruments above the steering column. On their right are the switches for the twin Marchal lamps, and the useful overdrive cut-out switch. This is an unusually well-equipped used car

Ford Consul II DE LUXE

The mascot in the centre of the grille identifies the de luxe model

FORD are now well into the fifth year of production of the Consul, and it says much for the soundness of the car's original conception that no major alterations have had to be made in the appearance or in the car's mechanical specification. Instead, the policy has been to avoid change merely for its own sake, and to concentrate instead on worthwhile detail improvements affecting comfort, convenience and the standard of the finish. In these respects the car has advanced a long way from the original version which was introduced in 1956: yet the basic price of the standard model has risen by only £25, and that of the de luxe Consul is the same as when the model was first added to the range for the 1957 London Motor Show.

It takes only a short time at the wheel for the driver to feel entirely at home with the Consul. The windscreen is broad and deep, and although the screen pillars are not slim, they are brought well back out of the way of forward vision. The seating position sets the occupant at ease and allows his full concentration to be directed to driving the car, unhindered by any awkwardness or lack of comfort. Noticed at once is the remarkable lightness of all the driving controls, and their smoothness. This really is a car which leaves the driver unfatigued after many hours at the wheel.

The engine is an over-square four-cylinder unit of just under 1¾ litres, which gives to the car an adequately lively performance without fuss. This is creditable, for the Consul is quite a large car—a true six-seater with generous luggage accommodation.

Starting is always immediate, and although the choke was necessary for the first start of the day in mild weather, it could be released entirely after a few seconds' warming-up, and the engine would then pull at its best almost immediately. Stalling does not occur at traffic halts in the intermediate period before normal running temperature is reached. There is no provision for a starting handle.

In the indirect gears the engine is reasonably quiet. There is certainly not sufficient engine noise to be troublesome or in any way obtrusive; on the other hand, a degree of power roar is heard when the upper part of the rev range is used. This does impose some slight restriction on the happy cruising speed which may be sustained on the open road. The Consul is happy at up to 70 m.p.h. Speed in top gear per 1,000 r.p.m. is 16·8 m.p.h., so that engine revs at 70 m.p.h. are few more than 4,000, and the power unit is not being overtaxed at this speed. The best maximum recorded in the test (with a following wind) was 80 m.p.h., corresponding to 4,750 r.p.m., which although above the point of peak power should still not be too fast for this short-stroke unit. At low speeds the engine is flexible, and will pick up smoothly from 15 m.p.h. in top gear, and from little more than a crawl in second.

In keeping with the effortless character of the car, and its ease of driving control, is the three-speed gearbox. On a 1 in 3 gradient the car refused a restart with the weight of two people on board, but except for this kind of severe test the choice of gear ratios proved to be excellent. Thus bottom gear takes the car from a standstill to 20 m.p.h., and proves very useful in traffic, when the Consul can usually be among the first cars away from any hold-up. In normal driving a change from second to top gear is usually made at about 45 m.p.h., and this middle ratio provides a useful range for rapid acceleration in main-road overtaking. If necessary, second gear can be held to well over 50 m.p.h., but there is no performance gain from delaying the change-up.

With a three-speed unit, of course, it is easier to provide a good steering column change for the gearbox, and that on the Consul is particularly pleasing; it will quickly reassure those who may have been discouraged by some of the early and less satisfactory four-speed column changes of various makes. The gear lever is light to operate and precise in its movement. Synchromesh is powerful on second and top gears, but is not provided—alas—on bottom. If there is any respect in which the Consul leaves room for improvement, it is this; sychromesh would eliminate the need to double-declutch on the many occasions when it is necessary to en-

Spring loading opens the luggage locker lid automatically when the button is pressed. The main part of the locker floor is rubber-covered, and the boot itself is large, with many corners for small luggage

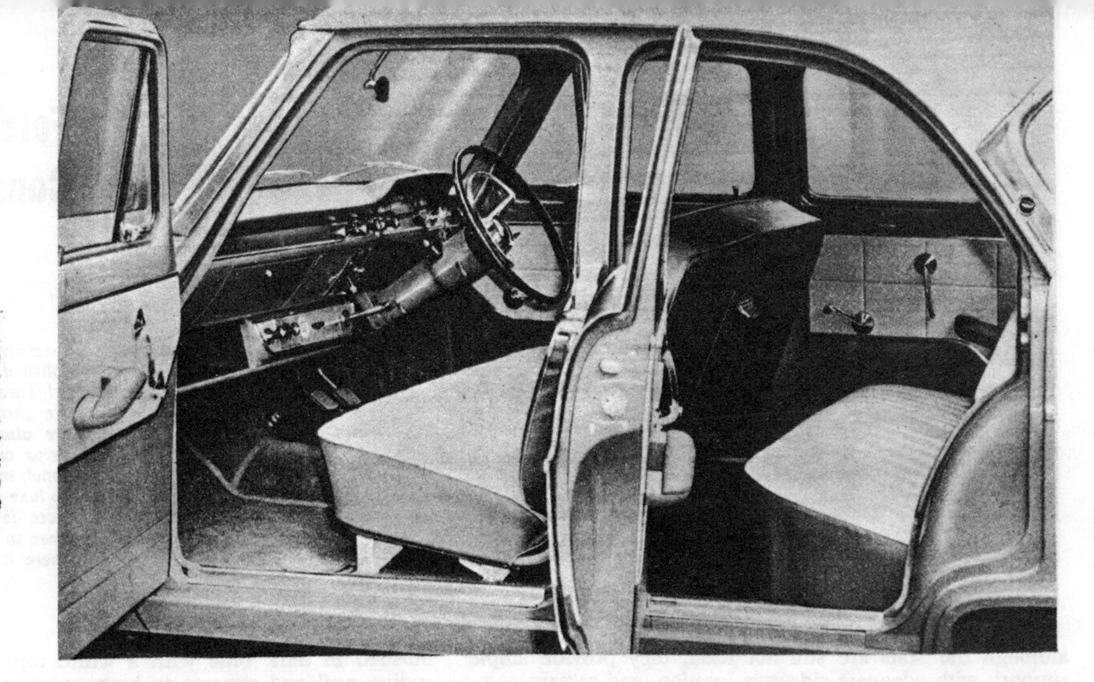

Both doors open wide, and access to front and rear compartments is easy. Ample space is provided for oddments in a lockable facia compartment and a roomy parcels shelf; there is also a map pocket on the trim-panel to the driver's right

gage bottom gear while the car is still rolling. In second gear a short spell of roughness was noticed at about 30 m.p.h. when accelerating hard, possibly resulting from some out-of-balance or maladjustment in the gearbox of the test car. There was also vibration of the gear lever at tickover in neutral.

Hydraulic control is used for the clutch and its action is extremely smooth. Pedal pressure required to operate it is light. Rather more clutch slip than usual is necessary when taking up the drive from rest: if the pedal is released rapidly before the car has a chance to get under way, too much engine power is absorbed, and the revs fall excessively. Also the clutch pedal has to be kept depressed for a second or two before engaging bottom or reverse gears. Too early movement of the gear lever into these positions results in noise from the gearbox.

Most noticeably light of all the controls is the steering, which calls for little effort at manœuvring speeds and becomes "featherlight" when the car is on the move. The control is precise, and on a straight road there is no wander at speed. This does not hold true on a windy day, however, because cross-winds have considerable effect on the directional stability, and appreciable steering wheel movements have to be made to correct deviations. Even on rough surfaces no wheel shocks are transmitted through the steering.

When cornering, the lightness of the steering is again appreciated, and the car can be pulled sharply round at speed with confidence, and without excessive effort. Indeed, the ability to make a violent swerve by a quick movement of the steering wheel could be a real safety factor in an emergency. A fair amount of roll occurs when corners are taken at speed, and on a winding road the occupants have need for all the available sideways location afforded by the seats. Tyre squeal is remarkably absent even when cornering to the limits of adhesion, and the balance of the car gives confidence to the driver, the mild degree of understeer being readily counteracted by acceleration. This is in hard driving, of course; in normal use the handling of the car is unobtrusively reassuring. On wet roads the Consul is notably resistant to skidding, and even when a slide is provoked the car remains easy to control.

Brake pedal pressures do not need to be high to produce quite urgent braking, and the maximum deceleration available was obtained with only 75lb pedal pressure; harder braking may result in wheel lock. For normal driving the brakes are amply powerful, and repeated violent use did not result in fade. When a full complement of passengers is on board slightly heavier pressure on the brake pedal becomes necessary, but here again the brakes are fully up to the job. The hand brake, however, will only just hold the car on a 1 in 4 gradient, and its lever is not as convenient to reach and operate as the remainder of the Consul's driving controls.

One of the best features of this car is its suspension. A sudden dip or wave in the road surface may sometimes provoke a slight lurch at the rear of the car, but this is the only fault, and in other respects the ride is considerably above average in comfort and freedom from pitch. Quite major road surface irregularities pass almost unnoticed, and on severe potholes the wide range of vertical wheel travel permitted by the suspension—particularly by the MacPherson independent layout at the front—is appreciated. When unmade roads are taken fast it is easy to see at once why this car, and its partner—the six-cylinder Zephyr—have done well in export markets where colonial conditions are the rule rather than (as in this country) the exception. Severe bumps occasionally produce a firm rebound movement at the front, but the dampers are powerful and prevent this from building up, and the bump stops absorb well the most severe shocks. In more normal conditions the ride is noticeably soft without feeling loose or giving rise to any tendency for the car to "float." Commendably little road wheel noise is transmitted to the body over any surfaces.

Much improvement has been made to the seat comfort of the Consul since the introduction of the first model, and

For owners who intend to carry out their own service on the Consul, under-bonnet accessibility is a delight, and all components are unobstructed. The dipstick is sensibly placed at the front of the engine

Ford Consul II . . .

Extensive use is made of polished metal in the exterior brightwork, but the bumpers are chrome-plated. There are also horizontal chrome strips on the number plate panel, marking this car as a de luxe model. The number plate is hinged, and folds down to reveal the fuel filler. There is no reversing lamp

although the seats are still not ideal, they provide ample support, with adequate sideways location, and remain comfortable on long journeys. The fault noticed is that there is rather too much springiness, which accentuates any slight vertical movement of the car. Softer, more "dead" upholstery would be better. A useful range of fore-and-aft adjustment of the front bench seat is available, and even a tall driver is able to provide himself with ample legroom for operating the pedals in comfort. In the rear compartment the occupants are by no means cramped even when the front seat is set at the limit of its rearward travel.

A centre folding armrest is provided in the front seat on this de luxe version of the Consul, and there are armrests on all four doors. At an extra cost of £3 10s plus £1 9s 2d purchase tax, a folding armrest for the rear seat may be specified. One of the most significant differences between the de luxe and standard models of the Consul is, of course, the use of leather (or nylon weave, to choice) for the upholstery. Other de luxe features are twin Windtone horns with powerful notes, operated by a semi-circular horn ring (instead of a central button); provision of a windscreen washer and cigarette lighter, and a vanity mirror on the rear of the left sun vizor and the fitting of carpets instead of moulded rubber floor mats in the front compartment. There are also exterior styling differences, of which the most important is the choice for the roof of a colour different from that of the bodywork. The test car was

The instrument layout is straightforward and functional, but the new rectangular speedometer is not as clear to read as the earlier semi-circular instrument. Durable-looking carpets cover front and rear floors

finished in dark blue with a white top, which suits the styling well and appears to be popular among owners.

The de luxe specification does not include a heater, but one was fitted to the test car. It warms the interior within a short distance of starting from cold, and is generously powerful. Improvements could be made, however, in the controls; the present arrangement makes it difficult to select an intermediate position to provide a gentle degree of warmth in mild weather, and prevents use of the rather noisy fan except when full cold or hot positions are selected —and then only at maximum fan speed: there is no rheostat control. The hinged quarter lights have a lip to prevent water running from them into the car in wet weather; and on a hot day they can deflect a pleasant draught of cool air to the interior when swivelled through more than a right-angle.

The much-criticised vacuum-operated windscreen wipers are continued. In their favour, they are self-parking and have a variable speed control on the facia. They clear a large area of the screen. At low speed settings they sweep regularly, but when the control is turned full on, the wipers become considerably influenced by engine speed and throttle opening. Their action is noisy, and the release of suction can be heard on each stroke.

Included in the overall fuel consumption figure of 24·3 m.p.g. for the entire test mileage are many miles in heavy traffic, performance measurement and runs at high average speeds. Little restraint is necessary in driving to improve the fuel consumption to a figure of 27 or more m.p.g.

Most significant of recent changes are those in the interior of the Consul. The new facia layout is considerably more attractive in appearance, and has the advantage that a padded facia top is provided to reduce injury in the event of a collision. Over the speedometer and instrument nacelle there is a deep hood which eliminates windscreen reflections at night, but makes the instruments sometimes difficult to read precisely by day. The fuel gauge for the 10½-gallon fuel tank is now of the same type as that fitted to the New Anglia, and gives a steady "dead reckoning" reading, good enough to justify more precise marking of the instrument than the familiar and rather vague E and F indications. The ammeter of the previous models is now replaced by a coolant thermometer—a useful improvement. Matching warning lights for the ignition, oil pressure, and head lamps' main beam, and separate left-and-right tell-tales for the flashing indicators, are retained.

To the right of the instrument nacelle is the lighting switch, which includes a rheostat control for the panel illumination. The switch cannot be confused with other controls, but it is not always easy to pull the switch just far enough to turn on the side lamps without going through to the head lamp position, or in the other direction to extinguish the head lamps without unintentionally turning off the side lamps as well. In the modern style, all the switches have pictorial marking to indicate their functions. The spread and penetration of the head lamps are adequate for the potential performance. With the panel lamps switched on, the rim of the cigarette lighter glows, to mark its position. For owner mechanics, there are 14 grease points for attention every 1,000 miles.

The new black steering wheel is well shaped and comfortable to hold. On the left of the facia the lockable glove compartment is still provided, and the useful parcels shelf below it now has a deep lip to prevent the contents from falling out on acceleration. A drawer-type ashtray is incorporated in the grille below the facia; its knob matches the cigarette lighter on its left. The sun visors are soft and well padded, and would not cause injury to the occupants in an accident. Other changes in the Consul over the past four years are not all immediately visible: for example, the fuel tank drain plug has been removed from its vulnerable position in the centre of the tank floor, where it was easily damaged on very rough roads.

This Ford Consul impresses by its simple honesty. While making no pretence to do more than serve as an efficient, unusually roomy family car, it does in fact offer more, for it has a vigorous performance, its driving ease is above average and its construction is particularly sturdy. It is also economical when not driven too hard. For a car which offers so much, the price is very reasonable.

FORD CONSUL II DE LUXE

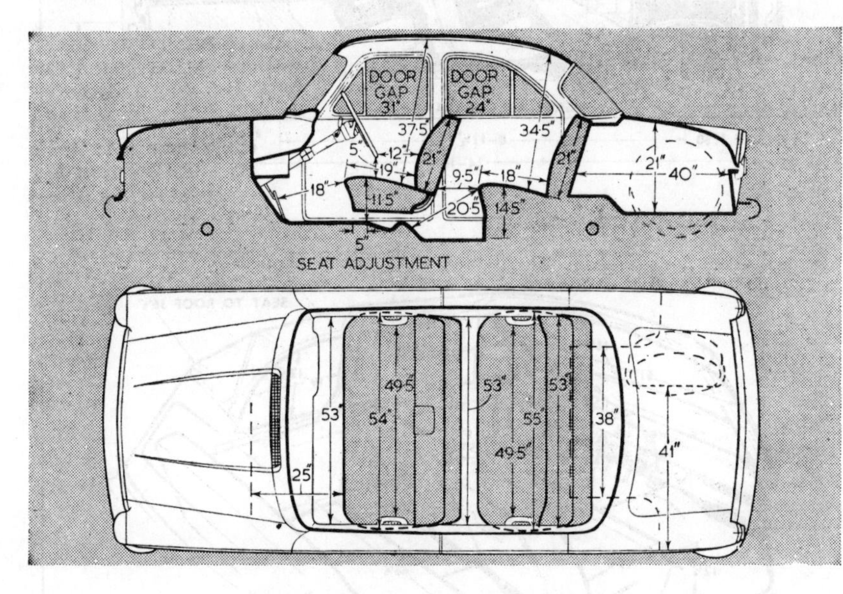

Scale ¼in. to 1ft. Driving seat in central position. Cushion uncompressed.

PERFORMANCE

ACCELERATION TIMES (mean):

Speed range, Gear Ratios and Time in Sec.

m.p.h.	4·11 to 1	6·75 to 1	11·67 to 1
10—30 ..	—	6·4	—
20—40 ..	11·3	7·1	—
30—50 ..	13·4	10·4	—
40—60 ..	17·5	—	—
50—70 ..	27·2	—	—

From rest through gears to:

30 m.p.h. ..	6·7 **sec.**
40 " ..	10·3 "
50 " ..	16·5 "
60 " ..	27·6 "
70 " ..	48·4 "

Standing quarter mile 22·7 sec.

MAXIMUM SPEEDS ON GEARS:

Gear		m.p.h.	k.p.h.
Top ..	**(mean)**	76·3	122·8
	(best)	80·0	128·8
2nd	.. ..	56·0	90·1
1st ..	.. ..	29·0	46·7

TRACTIVE EFFORT (by Tapley meter):

	Pull (lb per ton)	Equivalent gradient
Top	173	1 in 13·2
Second..	340	1 in 6·5

BRAKES (at 30 m.p.h. in neutral):

Pedal load in lb	Retardation	Equiv. stopping distance in ft
25	0·20g	150
50	0·38g	79
75	0·79g	38

FUEL CONSUMPTION (at steady speeds in top gear):

30 m.p.h.	38·0 m.p.g.
40 "	36·3 "
50 "	31·7 "
60 "	25·8 "
70 "	19·9 "

Overall fuel consumption for 1,072 miles, 24·3 m.p.g. (11·9 litres per 100 km.).

Approximate normal range 22–30 m.p.g. (12·8–9·4 litres per 100 km.).

Fuel: Premium Grade.

TEST CONDITIONS: Weather: Dry, 15 m.p.h. breeze gusting to 25 m.p.h.

Air temperature, 48 deg. F.

Model described in *The Autocar* of 27 February, 1959.

STEERING: Turning circle:

Between kerbs, L, 35ft 5in.; R, 35ft 3in.

Between walls, L, 37ft 3in.; R, 37ft 1in.

Turns of steering wheel from lock to lock, 3·75

SPEEDOMETER CORRECTION: M.P.H.

Car speedometer	10	20	30	40	50	60	70	80	86
True speed	10	19	28	37	46	55	66	75	80

DATA

PRICE (basic), with de luxe saloon body, **£580.**

British purchase tax, **£242 15s 10d.**

Total (in Great Britain), **£822 15s 10d.**

Extras: Radio to choice.

Heater **£11** plus **£4 11s 8d** purchase tax.

Whitewall tyres **£4 10s** plus **£1 17s 6d** purchase tax.

Rear seat centre armrest **£3 10s** plus **£1 9s 2d** purchase tax.

ENGINE: Capacity, 1,703 c.c. (103·9 cu. in.).

Number of cylinders, 4.

Bore and stroke, 82·55 × 79.5 mm (3·25 × 3·13in.).

Valve gear: o.h.v., pushrods.

Compression ratio, 7·8 to 1.

B.h.p. 59 (net) at 4,400 r.p.m. (B.h.p. per ton laden 46·7).

Torque, 91 lb ft at 2,300 r.p.m.

m.p.h. per 1,000 r.p.m. in top gear, 16·82.

WEIGHT (with 5 gal fuel): 22·2 cwt (2,492 lb).

Weight distribution (per cent): F, 53·7; R, 46·3.

Laden as tested, 25·2 cwt (2,828 lb).

Lb per c.c. (laden), 1·7.

BRAKES: Type, Girling, F, two-leading shoe; R, leading and trailing.

Method of operation, hydraulic.

Drum dimensions: F, 9·0in. diameter; 2·5in. wide.

R, 9·0in. diameter; 1·75in. wide.

Swept area: 240 sq. in. (190 sq. in. per ton laden).

TYRES: 5·90-13in.

Pressures (p.s.i.): F, 24; R, 24 (normal) F, 28; R, 28 (fast driving).

TANK CAPACITY: 10·5 Imperial gallons.

Oil sump, 6 pints.

Cooling system, 18 pints (plus 1 pint if heate fitted).

DIMENSIONS: Wheelbase, 8ft 8·5in.

Track: F, 4ft 5in.; R, 4ft 4in.

Length (overall), 14ft 5·8in.

Width, 5ft 8·6in.

Height, 5ft 0·1in.

Ground clearance, 6·4in.

Frontal area, 22·6 sq. ft. (approx.).

Capacity of luggage space, 18 cu. ft.

ELECTRICAL SYSTEM: 12-volt; 45 ampère-hour battery.

Head lamps, double dip; 50–40 watt bulbs.

SUSPENSION: Front, independent, coil springs, anti-roll bar.

Rear, live axle and semi-elliptic leaf springs.

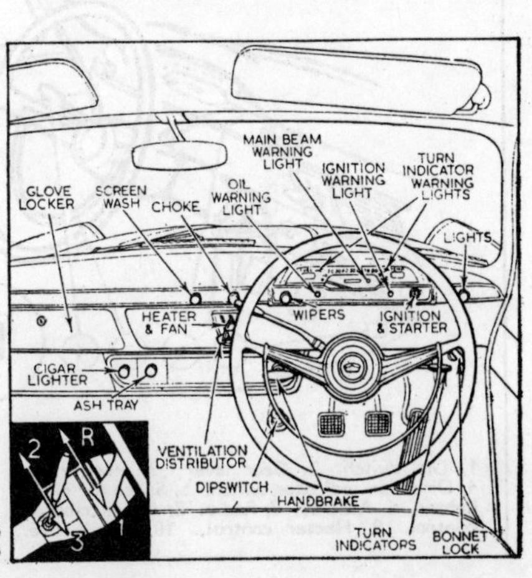

The Motor Road Test No. 17/61

Make: Ford. **Type:** Zephyr Convertible (with overdrive)

Makers: Ford Motor Co. Ltd., Dagenham, Essex.

Test Data

CONDITIONS: *Weather: Cool and dry with gusty 10-15 m.p.h. wind. (Temperature: 38°-44° F., Barometer 29.9 in. Hg.) Surface: Damp and dry concrete and tarred macadam. Fuel: Premium-grade pump petrol, approx. 96 Research Method Octane Rating.*

INSTRUMENTS

Speedometer at 30 m.p.h.	9% fast
Speedometer at 60 m.p.h.	7% fast
Speedometer at 80 m.p.h.	6% fast
Distance recorder	4% fast

WEIGHT

Kerb weight (unladen, but with oil coolant and fuel for approx. 50 miles)	25 cwt.
Front/rear distribution of kerb weight	56½/43½
Weight laden as tested	28¾ cwt.

MAXIMUM SPEEDS

Direct top gear.

Mean lap speed around banked circuit	88.3 m.p.h.
Best one-way ¼-mile time	90.9 m.p.h.

Overdrive top gear.

Mean lap speed around banked circuit	87.1 m.p.h.
Best one-way ¼-mile time	89.1 m.p.h.

"Maximile" Speed. (Timed quarter mile after one mile accelerating from rest.)

Mean of four opposite runs	85.0 m.p.h.
Best one-way time equals	87.4 m.p.h.

Speed in gears

Max. speed in overdrive 2nd gear approx.	80 m.p.h.
Max. speed in direct 2nd gear	64 m.p.h.
Max. speed in overdrive 1st gear	52 m.p.h.
Max. speed in direct 1st gear	37 m.p.h.

FUEL CONSUMPTION

(Overdrive top gear)

35.0 m.p.g. at constant 30 m.p.h. on level.
32.5 m.p.g. at constant 40 m.p.h. on level.
29.0 m.p.g. at constant 50 m.p.h. on level.
25.5 m.p.g. at constant 60 m.p.h. on level.
22.5 m.p.g. at constant 70 m.p.h. on level.
18.5 m.p.g. at constant 80 m.p.h. on level.
16.25 m.p.g. at maximum speed of approx. 87 m.p.h. on level.

(Direct top gear)

28.0 m.p.g. at constant 30 m.p.h. on level.
26.0 m.p.g. at constant 40 m.p.h. on level.
24.5 m.p.g. at constant 50 m.p.h. on level.
21.5 m.p.g. at constant 60 m.p.h. on level.
19.5 m.p.g. at constant 70 m.p.h. on level.
16.5 m.p.g. at constant 80 m.p.h. on level.
13.75 m.p.g. at maximum speed of approx. 88 m.p.h. on level.

Overall Fuel Consumption for 1,286 miles, 58.2 gallons, equals 22.1 m.p.g. (12.8 litres/100 km.)

Touring Fuel Consumption (m.p.g. at steady speed midway between 30 m.p.h. and maximum, less 5% allowance for acceleration) 24.5 m.p.g.

Fuel tank capacity (maker's figure) 10½ gallons

STEERING

Turning circle between kerbs:

Left	36½ ft.
Right	35 ft.
Turns of steering wheel from lock to lock	3½

BRAKES from 30 m.p.h.

0.97 g retardation (equivalent to 31 ft. stopping distance) with 65 lb. pedal pressure.
0.78 g retardation (equivalent to 38½ ft. stopping distance) with 50 lb pedal pressure.
0.37 g retardation (equivalent to 81½ ft stopping distance) with 25 lb pedal pressure.

TRACK :– FRONT 4'-5" REAR 4'-4"
OVERALL WIDTH 5'-8¾"
5'-0½" UNLADEN
22½" 13" 25¼" 15¾"
GROUND CLEARANCE 6"
SCALE 1:50
8'-11½"
14'-10½"
FORD ZEPHYR CONVERTIBLE

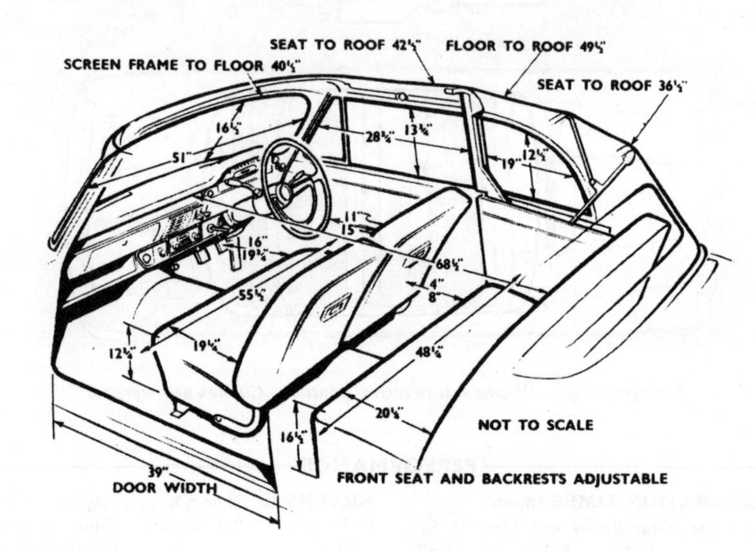

ACCELERATION TIMES from standstill

0-30 m.p.h.	4.8 sec.
0-40 m.p.h.	8.0 sec.
0-50 m.p.h.	11.3 sec.
0-60 m.p.h.	17.0 sec.
0-70 m.p.h.	25.8 sec.
0-80 m.p.h.	39.6 sec.
Standing quarter mile	20.6 sec.

ACCELERATION TIMES on Upper Ratios

	Over-drive Top	Direct Top	Over-drive 2nd	Direct 2nd
10-30 m.p.h.	—	8.0 sec.	—	5.0 sec.
20-40 m.p.h.	—	8.0 sec.	—	4.8 sec.
30-50 m.p.h.	14.9 sec.	8.8 sec.	7.4 sec.	5.6 sec.
40-60 m.p.h.	19.5 sec.	10.2 sec.	9.1 sec.	9.0 sec.
50-70 m.p.h.	24.0 sec.	14.4 sec.	13.6 sec.	—
60-80 m.p.h.	—	22.7 sec.	—	—

HILL CLIMBING at sustained steady speeds

Max. gradient on overdrive top gear	1 in 13.1	(Tapley 170 lb./ton)
Max. gradient on direct top gear	1 in 7.9	(Tapley 280 lb./ton)
Max. gradient on overdrive 2nd gear	1 in 7.5	(Tapley 295 lb./ton)
Max. gradient on direct 2nd gear	1 in 4.7	(Tapley 470 lb./ton)

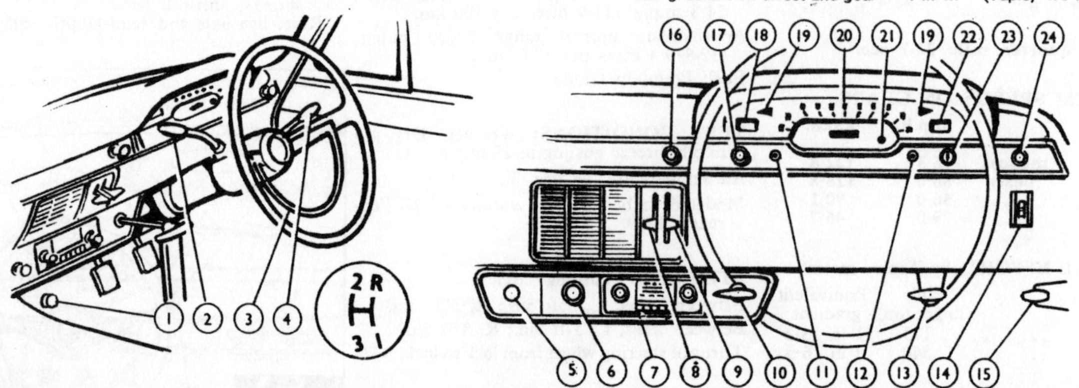

1, Dip switch. 2, Gear lever. 3, Horn ring. 4, Direction indicator switch. 5, Space for cigar lighter. 6, Ashtray. 7, Radio. 8, Air distributor control. 9, Heater control. 10, Handbrake. 11, Oil pressure warning light. 12, Dynamo charge warning light. 13, Overdrive. 14, Power-operated hood control. 15, Bonnet release. 16, Choke. 17, Windscreen wipers. 18, Fuel gauge. 19, Direction indicator warning light. 20, Speedometer. 21, Main beam warning light. 22, Water temperature gauge. 23, Ignition and starter switch. 24, Lighting switch.

—The Ford Zephyr Convertible (with Overdrive)

An attractive and roomy car, at a competitive price, the convertible Zephyr enables full advantage to be taken of sunny weather, though with the hood down a fairly leisurely pace has to be adopted to avoid draughts around the passengers.

Brisk Open-air Motoring for Five People

As a production model, the Mark II Ford Zephyr quietly celebrated its fifth birthday last month. It long ago ceased to be "news" and graduated to the select company of cars which, because they have pleased the customers, sell readily year after year without requiring more than very minor detail improvements. Whatever the merits of novelty, plenty of motorists who run a car for a big mileage before replacing it prefer well-tried designs of proven stamina, and well tried (in tough rallies as well as in ordinary service) the 1961 Ford Zephyr certainly is. In the Ford tradition of value for money, the convertible model which we tested recently has no rival at a comparable price in respect of the open-or-closed amenities, five-seat roominess and near-90 m.p.h. performance which it offers.

In convertible form the Zephyr has only two doors, whereas the equivalent saloon has four, and the well into which its roof can fold out of sight has been made by reducing both the length and breadth of the rear passenger compartment. It is probably fair to say that convertible bodywork "costs" the Zephyr one passenger seat, but this remains a practical five-seater with generous luggage capacity, taking three people in the front compartment (where there is only a very modest "hump" over the gearbox) and two more passengers in an easily-reached rear seat. Back seat kneeroom shrinks if a tall driver slides his seat well back, but footroom below the front seat remains ample—the limitation on rear-seat kneeroom arises partly because a driver in this car enjoys an exceptionally wide range of seat adjustment.

As a "driver's car" the Zephyr retains good all-round vision in its drop-head form, with quite reasonably slender windscreen pillars and only slight blind-spots in the rear quarters. Various members of the staff criticized an armrest on the right-hand front door as obstructing their arm movements whilst driving, and some comments about clutch operation being rather hard work were as much due to the height of the clutch and brake pedals above the floor as to the effort which had to be applied to them.

For a $2\frac{1}{2}$-litre car to be timed at 90 m.p.h. with very slight help from the wind is no longer remarkable, but a fine surge of top-gear acceleration anywhere in the speed range (from 20 to 40 m.p.h. in only 8 sec., or from 50 to 70 m.p.h. in 14.4 sec., without touching the gear lever) still makes the Zephyr a brisk-moving car on give-and-take roads. The smooth and apparently tireless engine was always an instant starter from cold (our test took place in quite cool spring weather) but needed the choke kept in partial use for about a mile after being started up if it was not to stall at traffic checks or spit back when the accelerator was first depressed.

The extremely smooth and willing engine is easily accessible for routine maintenance, as are the accessories distributed around the large under-bonnet compartment.

In Brief

Price (including overdrive, disc front brakes and power-operated roof as tested) £841 10s. plus purchase tax £351 15s., equals £1,193 5s.

Price with manual roof folding, no overdrive and drum brakes (including purchase tax) £1,028 4s. 2d.

Capacity	2,553 c.c.
Unladen kerb weight ...	25 cwt.
Acceleration:	
20-40 m.p.h. in top gear	8.0 sec.
0-50 m.p.h. through gears	11.3 sec.
Maximum direct top gear gradient	1 in 7.9
Maximum speed	88.3 m.p.h.
"Maximile" speed ...	85.0 m.p.h.
Touring fuel consumption	24.5 m.p.g.

Gearing: 18.45 m.p.h. in top gear at 1,000 r.p.m. (overdrive, 26.4 m.p.h.); 35.2 m.p.h. at 1,000 ft./min. piston speed (overdrive, 50.3 m.p.h.).

The Ford Zephyr Convertible

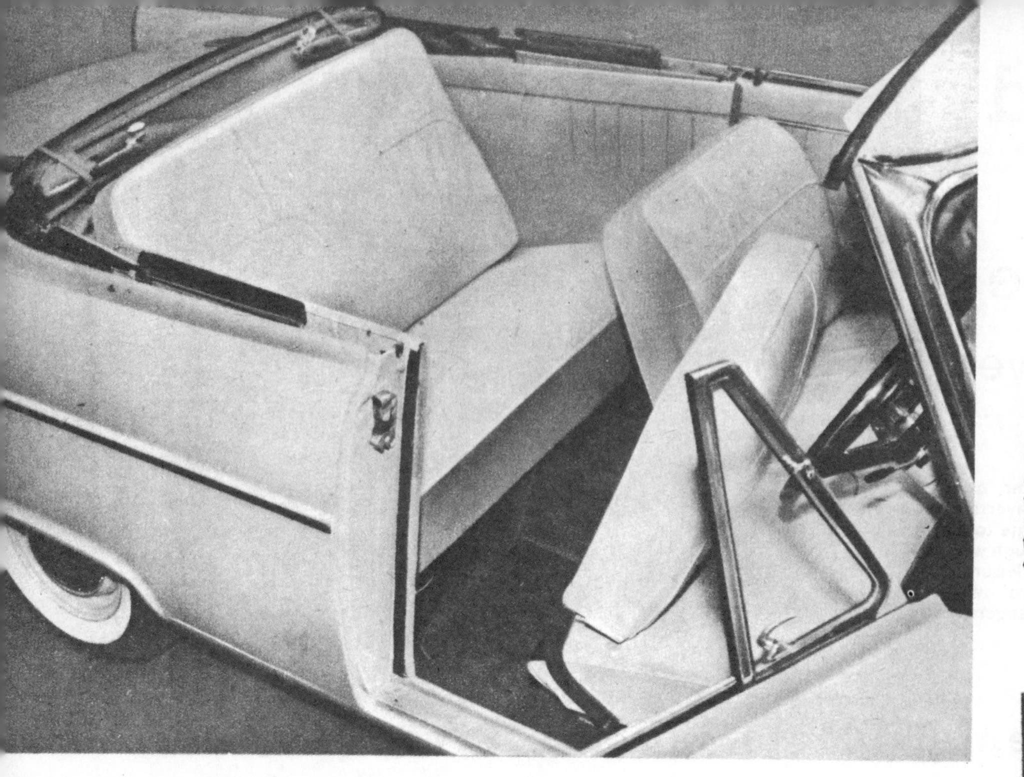

Tipping squabs which tilt inwards as well as forwards make access to the rear seat easy; the front seat can carry three people and has a wide range of adjustment. When folded, the hood is normally concealed by a cover, not shown in this picture.

The driving position proved comfortable for drivers of widely different stature, and visibility from the driving seat is good, but clutch and brake pedals are set rather high from the floor.

Our test car was fitted with the optional overdrive (accounting for about £60 of its tax-paid price) which, so long as the lock-out control has not been pulled, engages whenever the accelerator pedal is released at a speed above 32 m.p.h., and disengages when the car slows below 27 m.p.h. or when the driver applies "kick-down" pressure to move the accelerator pedal beyond the full-throttle position. As an aid to really fast cruising the overdrive was frustrated by excessive wind noise, which above 75 m.p.h. would hardly allow the engine, the radio or anything else to be heard; as an aid to fuel economy, overdrive undoubtedly helped this car to come within 1 m.p.g. of the overall consumption recorded in our last Road Test Report on the lower-powered Ford Consul, and let the Zephyr show slightly better cruising fuel consumptions at steady speeds than could the non-overdrive Consul.

Keen drivers, our staff seldom like three-speed gearboxes, especially those lacking synchromesh on 1st gear, but this Zephyr with quiet gears, an easy surge of power and the optional overdrive left little to be desired. Although bottom gear has no synchromesh, it is very easy to engage smoothly and silently at low speeds, and provides uninterrupted 0 to 30 m.p.h. acceleration when required; middle gear which has powerful (if at times slightly obstructive) synchromesh is useful up to 55 m.p.h. or so. Some of our staff have an incurable dislike of the semi-automatic overdrive and keep it permanently locked, whereas others use it all the time, running mostly in top and overdrive top gears on the open road or in 2nd and overdrive 2nd ratios around town: when the overdrive is unlocked, the car will free-wheel at speeds below about 27 m.p.h., and it is then possible to change gear without using the clutch. Perhaps owing to some characteristic of the flexible engine mountings, neither the take-up of drive from free-wheeling at low speeds nor the engagement of overdrive with the throttle closed were always as smooth as might have been wished.

A disadvantage of the high 1st gear (which is so advantageous in ordinary driving) is the inability to restart on any gradient steeper than about 1 in 4 without punishing the clutch severely. Drivers who like to move gently away from rest without much throttle also found it easy to stall the engine (which seems to have a very light flywheel) even on level road, although intentionally brisk starts were very brisk indeed and apt to be accompanied by wheelspin.

Riding comfort in this car attains about the standard which is nowadays regarded as "average" without calling for any higher praise. Rear-seat riding comfort is about equal to that in the front, where rather bouncy seat cushions seem to offset the advantage of an inter-axle position, but this is not a car in which a passenger can make very legible use of pencil and paper whilst being driven. There is a moderate but not excessive amount of roll on corners, the car showing a consistent (and never excessive) degree of understeer, and a driver is not liable to get any nasty surprise if he slightly mis-judges a corner or has to swerve suddenly to avoid other traffic.

Although the spare wheel is housed in the boot and a little room is taken up by the hood well, there remains really ample space for family luggage.

On the low geared side and not providing an especially good lock, the steering is light and mechanically precise, although at speed on cambered roads rather more inherent "sense of direction" would be welcomed.

Whilst its strength has been amply proved by now, the stiffness of the structure beneath this convertible body is not equal to that of the steel-roofed Zephyr saloon. The relatively unsupported body sides can shake on rough road, especially if the roof is folded but the sturdily framed windows have not been lowered, and at times some vibration of the steering column was also evident during fast driving.

Some extra bodywork vibration will be regarded by many people as a minor disadvantage when balanced against the pleasure of being able to enjoy fresh-air motoring. It is the work of very few moments to furl the front half of the roof into a "coupé de ville" position, and switch-controlled power operation then does the rest of the job of hood folding or of subsequent re-erection: fitting the cover tidily over the folded hood is a slightly tedious job which many owners will only undertake when the weather is truly "set fair."

In Britain's climate the "coupé de ville" position of the roof is perhaps more used than any other, since it provides fresh air and a view of the sky but does not prevent the interior heater being effective in warming the driver's feet. Fully opened, there is inevitably a fair amount of draught inside as roomy a body as this five-seater, and, as few people nowadays wear heavy motoring coats, driving with the roof and all windows lowered is likely to be done mostly at leisurely holiday speeds.

All the amenities of a modern saloon car are provided inside this convertible, even to sun visors and an interior light (with door-operated courtesy switches) mounted above the windscreen. Interior heating is an optional extra, which worked well on the test model: a radio with five push-buttons is of a design produced especially for Ford cars, and this optional extra also worked well at normal speeds although in motorway conditions it was drowned by the excessive wind noise. On the facia, a lockable glove box is supplemented by a shelf beneath it. The two big doors cannot be opened by the interior handles if the locking buttons are depressed, so that a child should be safe even in the front of this two-door body, but either door can be locked or unlocked with the key from outside the car.

Neat and weatherproof, the hood offers saloon-car snugness in cold weather but causes considerably more wind noise than the saloon.

Our test car had the optional disc front brakes, which we understand are chosen by a majority of buyers, the price quoted on page 549 including both these and the overdrive. The instant and smooth response to moderate pressures on the brake pedal, at high speeds or low, was most reassuring. This may be regarded as adding the finishing touch to a sensibly priced modern version of what novelists used always to describe as a "high-powered touring car."

Specification

Engine

Cylinders	6
Bore	82.55 mm.
Stroke	79.5 mm.
Cubic capacity	2,553 c.c.
Piston area	49.74 sq. in.
Valves	o.h.v. (pushrods)
Compression ratio	7.8/1 (optional 6.9/1)
Carburetter	Zenith pump-type downdraught
Fuel pump	Mechanical, incorporating vacuum pump
Ignition timing control	Centrifugal and vacuum
Oil filter	AC or Tecalemit full-flow
Max. power (net)	85 b.h.p. (90 gross)
at	4,400 r.p.m.
Piston speed at max. b.h.p.	2,300 ft./min.

Transmission

Clutch	8½ in. single dry plate
Top gear (s/m)	3.90 (overdrive, 2.73)
2nd gear (s/m)	6.40 (overdrive, 4.48)
1st gear	11.08 (overdrive, 7.76)
Reverse	15.06
Overdrive	Borg-Warner semi-automatic with 0.7/1 ratio
Propeller shaft	Single-piece open
Final drive	Hypoid bevel
Top gear m.p.h. at 1,000 r.p.m.	18.45 (overdrive, 26.4)
Top gear m.p.h. at 1,000 ft./min. piston speed	35.2 (overdrive, 50.3)

Chassis

Brakes: Girling hydraulic with vacuum servo, disc at front and drum at rear on test model.

Brake dimensions: Front discs 9¾ in. dia.; rear drums 9 in. dia. × 1¾ in. wide.

Friction areas: 81.1 sq. in. of lining (20.6 at front and 60.5 at rear) working on 299 sq. in. rubbed area (200 at front and 99 at rear)

Suspension:
- Front: Macpherson-type independent by telescopic damper struts, coil springs, anti-roll torsion bar and lower wishbones.
- Rear: Rigid axle and ¼-elliptic leaf springs.

Shock Absorbers:
- Front: Telescopic incorporated in suspension struts.
- Rear: Armstrong lever-arm hydraulic.

Steering gear: Recirculatory-ball bearing worm and nut.

Tyres: 6.40-13, tubed or tubeless to order.

Coachwork and Equipment

Starting handle	None
Battery mounting	Alongside engine on right
Jack	Bevel-geared bipod type
Jacking points	External sockets under each side of body

Standard tool kit: Jack, wheel brace and hub cap remover.

Exterior lights: 2 headlamps, 2 sidelamps, 2 stop/tail lamps, 2 rear number plate lamps.

Number of electrical fuses	Two

Direction indicators: Self-cancelling flashers, white front and amber rear.

Windscreen wipers: Self-parking twin blade, vacuum operated with engine driven booster pump.

Windscreen washers	Optional extra
Sun visors	Two

Instruments: Speedometer with decimal total distance recorder, fuel contents gauge, coolant thermometer.

Warning lights: Dynamo charge, oil pressure, headlamp main beam, turn signals.

Locks:
- With ignition key: Ignition/starter switch, either door, glove box, luggage locker.
- With other keys ... None

Glove lockers: One on facia with lockable lid.

Map pockets ... None

Parcel shelves: Below passenger's side of facia; well behind rear seats can accommodate parcels when hood is not folded down.

Ashtrays: One on facia, two behind front seat.

Cigar lighter ... Optional extra

Interior lights: One above windscreen with courtesy switches.

Interior heater: Optional extra fresh air heater and screen de-mister.

Car radio ... Optional extra

Extras available: Leather upholstery, heater, radio, white-wall tyres, overdrive, servo-assisted disc brakes, and full range of accessories.

Upholstery material: P.V.C. plastic (leather as optional extra).

Floor covering	Carpets
Exterior colours standardized	Ten

Alternative body styles: Four-door saloon and estate car; also Zodiac saloon, estate car and convertible.

Maintenance

Sump: 7 pints, plus 1½ pints in filter, S.A.E. 20 or 20W for temperate summer and winter weather.

Gearbox	2½ pints, S.A.E. 80 EP gear oil
Rear axle	2½ pints, S.A.E. 90 hypoid gear oil
Steering gear lubricant	S.A.E. 90 EP gear oil

Cooling system capacity: 22½ pints (2 drain taps).

Chassis lubrication: Every 1,000 miles by grease gun to 12 points, and by oil gun to 2 points.

Ignition timing	8° before t.d.c. static
Contact-breaker gap	0.014-0.16 in.
Sparking plug type	14 mm. Champion N8
Sparking plug gap	0.032 in.

Valve timing: Inlet opens 17° before t.d.c. and closes 51° after b.d.c.; Exhaust opens 49° before b.d.c. and closes 19° after t.d.c.

Tappet clearances (hot): Inlet and exhaust 0.014 in.

Front wheel toe-in	1/16 to 1/8 in. unladen
Camber angle	½° to 2¼° unladen
Castor angle	0° to 1¼° unladen

Steering swivel pin inclination: 3½° to 4½° unladen.

Tyre pressures	Front and Rear 24 lb.
Brake fluid	Enfo

Battery type and capacity: 12 volt, 57 amp. hr.

A USED CAR
ROAD TEST

1958 FORD CONSUL

ALTHOUGH lighter, less expensive and yet offering the same roominess as its bigger brother, the Ford Consul has not, in the Union that is, enjoyed the same measure of popularity as the Ford Zephyr. One of the reasons for this preference for the larger car is the fact that both cars use about the same amount of fuel. The economical disadvantages of owning a Zephyr instead of a Consul is that the annual licence is higher and repairs to the engine will be more costly, but then it is a far superior performer. The same firm who recently lent us a Zephyr for road testing gave us a Consul with just over 44,000 miles on the clock. Their price for the 1958 model was R990 — it cost R1,568 when new. Thus the depreciation for nearly three years of use is R578, which is average in the case of the smaller type of car.

Bodywork and Upholstery

When we went to collect the Consul at the dealer it stood on the showroom floor and was beautifully prepared for prospective customers. The cream paint work was in good condition and it could be seen that it was the original, although a certain amount of touch-up work had been done in the region of the doors, bonnet and boot. The damage to the paint had apparently only been minor scratches collected in the course of normal use. After close inspection of the body panels it was revealed that these had never received the attentions of a panel beater. All the doors closed firmly, although the pressure needed in the process was rather excessive. The windows of all the doors worked smoothly and were free of cracks. The window on the driver's side tended to rattle on rough surfaces, this being due to slightly worn channelling. The windscreen and rear view window of the Consul were free of cracks and the rubber sealing was in good condition and perfectly waterproof.

Both boot lid and bonnet closed firmly and easily and it could be seen that the underside of both these components had been resprayed.

A glance at the interior of the car showed that the previous owner or owners had cared well for it. The interior of the car had obviously also been spruced up in the dealer's workshops, but this rarely hides previous damage. The hood-lining was neat and had never been removed for repairs to the roof. Upholstery and piping in the region of the doors was neat and in a good state of preservation. Both front and rear seats were neat and the springs in these firm.

Electrical Components

All the electrical components fitted to the Consul were in good working order, the only exception being the battery, which either needed recharging or replacing as it almost refused to turn the starter motor when the headlights were switched on. The generator delivered the correct rate of charge and all the lights, at the front and rear, worked. The beams of the headlights were cast at the correct angles and matched the car's performance when driving after dark. The wiring under the bonnet was in good condition.

Instruments

The instruments of the Consul all functioned. The speedometer was correct at the 20 mark, but became progressively faster as the speed was increased. It was found that the fuel gauge was accurate but the needle stuck when the ignition was turned on and this component needed a hefty bang to release the reluctant needle. Both the warning light for ignition and the engine oil system came on at the correct intervals and the warning lights for the traffic indicators functioned correctly.

Dust- and Waterproofing

As far as we could judge the Consul seemed waterproof. Where the dust-proofing is concerned it can be said that the rubber sealing was still in good condition, taking into consideration that the car had covered more than 44,000 miles. A fair amount of dust seeped in through the bottom of the front doors and spread along the floor. This was rather disappointing, because the rear doors were almost perfectly dust-proof. After a dusty section of the test route had been negotiated a fairly thick layer was found in the boot.

Steering

There was a certain amount of play in the steering-box of the Consul, normal for a car which had covered over 40,000 miles. The whole assembly, however, was firm and there was no tendency to pull to either side. The alignment was correctly adjusted.

Engine

Mechanically the Consul's engine was in good shape, but it was in need of tuning. The bad state of tune accounted for the car's rather healthy appetite for fuel. The only noises heard from the engine were tappets which needed adjustment and a slight piston slap after it had been hot. The fact that the engine used no oil after 44,000 miles seems to suggest that the rings had been renewed. When the needle of the speedometer crept over the 60 mark the engine ran rough, which was due to the contact points needing attention or replacement. It was possible that the spark plugs also needed renewing. The bearings, both main and big ends, were firm and never made themselves heard. At the top end the valves were sound and there were no signs of compression leaks. It was noticed during the entire test run that the engine never exceeded its normal operating temperature.

Transmission

The transmission on the Consul was silent and the gears never jumped out of position. At the rear the differential was silent — a significant fact when buying a Consul as these cars are sometimes plagued with noisy diffs. The gear-change linkage operated smoothly, and the clutch did not slide.

Suspension

The front suspension of the Consul was firm and held the car's nose well on rough surfaces. There was a certain amount of wear on the kingpins, but the tie-rod ends were firm as were the shock absorbers. The car tended to wag its tail on rough surfaces due to weak shock dampers at the rear. Pivot pins in the front and shackles at the rear were firm.

Brakes

The braking system of the Consul was still in good condition and there was no tendency to pull to either side. In an emergency stop the brakes took squarely and the car stopped in a straight line. The tyres fitted to the Consul were practically new while the spare wheel had a new retread.

Performance and Ride

As was previously mentioned the Consul's performance could have been improved on by tuning the engine, but everything taken into consideration the car performed well. A glance at the car would have convinced even an inexperienced person that it had been well cared for, and this fact was further emphasised when the car was driven over rough surfaces, for the body was remarkably tight and free of rattles.

Summary

If the Consul's economy is improved it is the ideal car for the man seeking a medium-sized vehicle with comfortable riding qualities and is worth the R990 asked for it by the dealer.

MAKE, MODEL AND YEAR OF MANUFACTURE: 1958 Ford Consul saloon.
PRICE WHEN NEW: R1,568.
PRICE SECOND-HAND: R990.
INDICATED MILEAGE: 44,363.
WARRANTY: A.1.
FUEL CONSUMPTION: 21·85 m.p.g. at 57·18 m.p.h.
BRAKING EFFICIENCY (with car in neutral at 30 m.p.h.): 85 per cent. (equivalent to 36 ft. stopping distance).
PERFORMANCE:
(Through gears):
0 to 30 m.p.h.: 6·6 secs.
0 to 60 m.p.h.: 26·9 secs.
(In top gear):
20 to 40 m.p.h.: 11·7 secs.
30 to 50 m.p.h.: 13·7 secs.
40 to 60 m.p.h.: 16·0 secs.
(Car supplied by Holmes Motor Co. (Pty.) Ltd., Cape Town).

Badging and grille bar pattern identify this as a Zephyr Six — six-cylinder cars are 9ins longer than 'fours'

FIVE STAR TRAVEL

Ford's first all-new cars after World War II, the Consul and Zephyr models launched in 1950, have a growing following. Peter Williams outlines the appeal of these 'Five Star' cars

'Five Stars' — that was the marketing pun used to describe the salient features of the new Ford Consul and Zephyr in 1950, when Ford's first all-new cars since World War II were launched at the Earls Court Motor Show. The pun was used to describe the five areas where the Consul and Zephyr heralded 'firsts' for British cars: use of MacPherson struts, hydraulic brakes all round, 'oversquare' engines, between-the-wheels seating position, and monocoque construction.

Early monocoque

In reality the word monocoque is misleading, because the Mk1 Consul and Zephyr — as the Five Star cars came to be known — actually featured quite substantial chassis members welded to the main floorpan and extending forward to take the engine and front suspension. The all-new bodyshell was the work of George Walker and was evolved from Ford of America's own first all-new post-war saloon, which was thought to be too big and heavy to market in Britain in the way Dearborn products had been before the war. Front wings and the front panel were all bolt-on, and the structure ahead of the A-post differed in length on the four-cylinder (Consul) models and the six-cylinder (Zephyr) models.

One reason why Dagenham's designers chose monocoque construction was the development by an American Ford engineer, Earle MacPherson, of the independent MacPherson strut. At the rear, the Mk1 carried a live axle supported by leaf springs and lever-arm dampers. Steering was worm and peg, and drum brakes featured all round. Wheels were 13ins diameter shod with 5.90 × 13 tyres on the four, and 6.40 × 13 tyres on ½in wider rims on the 'six'. The extra inches for the bigger engine could be found in the wheelbase, 100ins for the four and 104ins for six.

An oversquare engine configuration was adopted with the four and six sharing major components. The engines were nothing special, but at the time were state-of-the-art units with overhead valve design and a short throw crankshaft. The four had a bore and stroke of 79.37mm × 76.2mm to give a capacity of 1508cc, and the two additional cylinders for the six with the same bore and stroke gave 2262cc.

The head featured siamesed inlet ports and the low compression ratio of 6.8:1 was necessary at the time due to the low octane petrol then used. The four developed 47bhp at 4400rpm and 74lb ft torque at 2400rpm, with corresponding figures for the six of 68bhp at 4000rpm and 108lb ft at 2000rpm, which gave both engines good low-down pulling ability.

Both four and six-cylinder cars featured three-speed gearboxes with a column mounted change which through final drive ratios of 4.625:1 and 4.375:1 respectively gave low overall gearing and also endowed the cars with good top gear acceleration. Both pull easily from 10mph in top and go on to maximum speeds of around 73 and 80mph respectively. During the Mk1's lifespan the final drive ratio was changed to 4.556:1 on the Consul and 4.444:1 on the Zephyr Six.

Initially the Five Star cars, called Consul and Zephyr Six (the latter so as not to be confused with the American V12 engined Lincoln Zephyr), were not particularly well-equipped, in keeping with the austere standards of the period. Upholstery was PVC (leather was an option), and both front and rear seats were devoid of a central armrest to accommodate six people. There was a cloth roof-lining on both cars, but the Consul had a rubber mat floor-covering as opposed to the carpeting of the Zephyr Six. A heater was initially an option on both cars.

It was not immediately apparent that the four and six-cylinder cars were different except for badging and front grille treatment, or that the sixes were actually 9¼ins longer, at 14ft 3¾ins. The Consul featured a rectangular grille with vertical slats, and the Zephyr Six a raised section design with horizontal bars.

Competitive pricing

When the Five Star cars went on sale in January 1951, the price of the Consul was £532: it competed against the Austin Devon at £537 with a 1.2-litre engine, the 1¼-litre sidevalve Hillman Minx at £543 and the Morris Oxford MO series which boasted a 1½-litre sidevalve engine for £573. Rivals to the £608 Zephyr Six were the 2.2-litre Morris Six at £718, the 2.2-litre four-cylinder Austin Hereford at £738 and the Standard Vanguard at £703 with a 2.1-litre four-cylinder engine. In short, the Fords had price on their side.

Production started at the beginning of 1951, and *The Autocar* published the first full road test of a Consul soon after: 'It is one of the outstanding cars produced since the war in the popular class and has handling qualities that would be acceptable on a car of any price.' The magazine also said that the Consul engine was smooth, lively and quiet.

The Autocar was similarly enamoured by the Zephyr Six when it tried the model towards the end of 1951. Summing up, Raymond Mays said: 'The Zephyr is a truly outstanding car in almost every respect. It corners like a racing car — better than some racing cars, in fact. Where road conditions permit, you can make maximum speed and cruising speed synonymous, and it takes this sort of medicine indefinitely.'

By this time, after only one year's production, the Mk1 received some minor improvements, the most notable being the replacement of the 'flat dash' for one which featured a full width parcel tray, an item which had come in for criticism from the motoring press.

Throughout the Mk1's life, more improvements to specification followed. For example, in 1951 a prototype convertible version was shown at Earls Court with actual production of converted saloons starting in 1952 at Carbodies of Coventry. But the model to cap the Five Star model range was not destined to arrive until the 1953 Earls Court Show, when a 'de luxe' version of the Zephyr Six called the

Winning the Monte Carlo Rally with a Zephyr in '53 gave Ford a useful push in the marketplace

Abbott of Farnham made estate versions — were they tank-like? — which are very rare now

Only around 30 convertibles (this is a Consul) survive

Left-hand drive dashboard layout of a '53 vintage Zephyr

'Zephyr Zodiac' was revealed. It incorporated the heater and leather upholstery listed as options from the start, boasted a two colour paint scheme and even more chrome, and was also further identified by spot, fog and reversing lights.

Inside, the Zephyr Zodiac featured a high level of equipment, but, more importantly, under the bonnet was a more powerful version of the six-cylinder engine with a new cylinder head to give a 7.5:1 compression ratio, and a higher power output of 71bhp at 4200rpm to take advantage of recently introduced high octane petrol. It cost £851 and some of the specification improvements were carried over to the Consul and Zephyr Six, including the 7.5:1 compression ratio for the latter as a no-cost option.

Luxury convertible

The luxurious Zephyr Zodiac was naturally available as a convertible too. Another variation of the Five Star theme was an estate conversion by E.D. Abbott of Farnham, and the model also benefited from performance tuning by a number of specialist companies.

The Mk1 also had an illustrious competition history, beginning with an outright win for the Consul in the 1952 Tulip Rally. Later in the model's life automatic and overdrive were gearbox options on the six-cylinder cars, and desirable ones at that.

When production ended in February 1956 the Five Star cars were replaced by the Mk2 or 'Three Graces' — a total of 406,792 had been produced with a break-down of 231,481 Consuls, 152,677 Zephyr Sixes and 22,634 Zephyr Zodiacs.

So what do you look for in a Five Star car? Like many early post-war British saloons, they are pretty cheap and despite 'classic' status nowadays are not terribly sought-after. As the six-cylinder cars are more in demand, decide first whether you want a Consul, a Zephyr/Zephyr, or a Zodiac. Both engine installations can keep up with the rest in modern motoring conditions, can seat six, take a fair amount of luggage and cruise well over long distances.

Prices are entirely dependent on condition: saloons are cheapest, the convertibles the most desirable. A tatty running Consul would cost from £250-400, with an unrestorable car not more than £100. Such a car is a worthwhile back-up because there are a lot of valuable spares on them. A mint Consul ought to be worth £750-1000, but the market rarely supports the higher price.

With the six-cylinder cars more in demand, a tatty Zephyr can fetch £400 — you don't see MoT failures because enthusiasts keep them going. The Zodiac version is another £50-100 premium on that figure. A Zephyr runner would be £600-700 and a similar Zodiac £700-800, with mint cars at £1000 and £1250 *plus* respectively. Concours cars have been valued at £2000, but it's doubtful if they would sell for that.

The desirability factor of convertibles means that runners will fetch more than £1000, but it is more than difficult to find one as there are believed to be only 25-30 left in the country. Good examples are not often for sale as people hang on to them, but one would fetch £2500 upwards. You are highly unlikely to be able to buy an estate built by Abbott of Farnham — there are only three known in the country!

Mechanically, the Mk1 'Big Fords' are simple cars, and once kept in good condition are not going to 'go' without warning. With modern engine and gearbox oils, the engine will last a long time, the 'four' 70,000 miles if treated carefully and the 'six' quite easily 100,000 miles.

Look for all normal points of deterioration when examining a Mk1. Blue exhaust smoke is the obvious sign of a worn engine and if the big ends are gone you will hear those, though don't mistake it with the 'tappety' nature of the engines. Generally, the engines are not sensitive to oil pressure, though a compression check might be a more useful guide.

The weakest point of these cars is the steering linkage. The 'box' will last either a given number of miles or years, say 50,000 miles or 25 years. A new one will be needed, and that is both expensive and difficult to find. To spot the problem, if the steering wheel is turned it will first be tight and then there will be a slack spot. If the steering linkage itself has gone then the wheel will require some turning before anything happens.

To solve it, the Consul Zephyr Zodiac Mk1 Owners Club has found a way of regenerating old ones and using parts from cars in breakers yards. At the heart of the problem is the drag link, so an original Ford item is used with adapted Triumph Herald track rod ends. This method has been devised by Ford specialist Ken Tingey as there are currently no replacements available for either the drag link or the front struts.

Second gear goes first in the three-speed gearbox. A 'crunch' when changing down from top to second indicates that the synchromesh is on its way out on second. Difficult parts in this area are oil seals, the rear pinion and rear axle oil seals, though rear axles are said to last a 'lifetime'. On the steering column change, the idler arm bushes are the only thing to watch, requiring replacement every 20,000-25,000 miles.

Like the steering gear, the club is investigating the re-manufacture of the various oil seals to their original

Zephyr Six 2262cc engine: note simple exhaust manifold

Consul's 1508cc four-cylinder had same bore and stroke

specification, while similar moves are expected for body panels. These are very rare and when 'new' old stock is found it is often with poor tolerance levels . . . in other words rejects!

Rust is inevitably a problem. It's at its most serious at the side of the strut towers on the inner front wings, and rudimentary chassis rust should be re-plated anywhere else that would result in an MoT failure. New outside sills are available, while the floor rots through just behind the driver's right heel and the passenger's left heel, and also just ahead of the rear wheel arch on both sides. Water in the jacking points is the cause, and it's not difficult to repair but will require professional help.

Some sample prices are £18 for each of the floor sections with jacking points, and £23 for the sills. Incidentally, cars which have been left standing for a long time are prone to the engine rings sticking. The best solution is to 'rock' them in gear. They are also prone to the clutch sticking if left standing for more than a year.

Join the club

As you would expect, the best advice to give to a potential Mk1 owner is join the club. Formed in 1977, it is really the *only* source of spares, though there are knowledgeable specialists for the model. Its full title is Ford Mk1 Consul Zephyr and Zodiac Owners Club, and the membership secretary is Neil Tee of 8 Park Farm Close, Shadoxhurst, near Ashford, Kent TN26 1LD. We are grateful to Neil for his help in compiling this article.

Annual subscription to the club is £9 plus a £2 joining fee for which you get a monthly newsletter with lots of worthy snippets, particularly on spares. Membership is now up to 300.

Two-tone paint and extra lights mark out the Zephyr Zodiac — contemporaries looked dated

Two of the star Five Stars at a club gathering are Neil Tee's Consul and Zephyr Zodiac rag-tops

MARKET UPDATE FORD MkII CONSUL, ZEPHYR & ZODIAC

We last profiled Ford's 'three graces' in November 1985. A lot has happened in the past four years, as Peter Simpson discovers.

When we last looked at the MkII Fords, we quoted a top price for a MkII Zodiac convertible of £3000, and said that a basket-case convertible could be found for 'from £200', although suggested spending 'at least £500' to get a more worthwhile project. As for the saloons, well we said that you could probably get a condition one Consul for £1200 and a Zephyr or Zodiac could be found for perhaps £300 more.

Those were the days! As anyone who studies the classifieds must be aware, you can treble most of these figures today at least. Almost overnight it seems that the MkII Fords have become highly popular, so much so in fact that, unless ridiculously overpriced, a good car will sell very quickly indeed. We therefore decided that it was time to look at the big late fifties Fords again to see what the 1990 buyers can expect for their money.

The range

As most readers will be aware, there were three models, the Consul, Zephyr and Zodiac, with the Consul as the base model, the Zephyr the next one up, and the Zodiac as the most luxurious of all. All were introduced

FORD MkII CONSUL, ZEPHYR & ZODIAC

in 1956 as replacements for the trend-setting MkI range, which was in many ways the predecessor of most of what Ford in Britain did until the late 1970s. From the start saloons and convertibles were available and, later that year, an estate car was added. Convertibles and estates were both saloon conversions, the former by Carbodies (of FX4 fame) and the latter by Abbotts of Farnham. An 'unofficial' MkI estate conversion was also undertaken by Abbotts, but the MkII was done with the factory's full approval and featured in Ford's advertising. As with many Fords, there was a mid-production facelift (in February 1959). There were numerous trim and detail changes including a redesigned dash and oblong speedometer, improved upholstery and different lights, chromed headlamp bezels being a good facelifted model identification. The most noticeable change though was a flatter roof panel. This led to a reduction in overall height and the new models were consequently known as lowlines. A significant mechanical change came in October 1960 when front disc brakes (with servo assistance) were offered as an option. They were standard from May 1961. The last MkIIs were made in April 1962. After the Consul Classic was introduced in 1961 the MkII Consul was called the Consul 375. Quite why is a mystery; certainly the figure 375 bears no relation to anything in the car's specification.

Changes, changes and more changes

So what's made the MkII Fords so popular recently then? Undoubtedly their evocative fifties styling has a lot to do with it; with classic motoring as fashionable as it seems to be at present they are just the thing to be seen in! Although they are particularly popular with young people, according to several of the people I contacted in preparing this feature the MkIIs are also sought-after by many older people who remember the cars from their youth; the sort who perhaps were 'Teds' at the time but couldn't afford a car of any kind, let alone a new or nearly new MkII! It's also interesting to reflect on how many more dealers are selling MkII saloons these days. Five years ago most were interested only in convertibles or possibly the odd exceptional six-cylinder saloon!

Probably the most important consequence of the increase in value enjoyed by these cars recently is that few Consuls, and hardly any Zephyrs and Zodiacs, are now scrapped. While good for the marque's survival in many respects this has one unfortunate side-effect; there are now fewer secondhand spares about. Certain mechanical components are shared with the MkIIIs which

A MkII Consul (left) and Zodiac (right). Zephyrs and Zodiacs had longer front wings – to make room for the straight-six engine. Both these cars were supplied for us by Worthing Carriage Company; the lowline Consul has sold, but if you're quick you may be in time for the highline Zodiac. It's a low (49,000) mileage car, on offer at £6,995.

A fine interior. Early MkIIs had all-metal dashes and instrument panels similar to the MkI range. Bench front seats were of course a feature of the era. Quite simple construction-wise, they are easier to re-cover than most.

helps, but no body panels are and these are still the main shortage area. In the time since we last looked at them the MkIIs have also suffered at the hands of that curse of the classic car scene, the numberplate dealer. Nowadays, because they're worth much more with the original plates, few MkIIs are separated from their original marks but up to three years ago many Consuls in particular were. There's no way you'll get an original mark back on an 'A' registered car other than by buying one through the trade. Consider doing this if you are unfortunate enough to own an 'A' registered Consul, and definitely do it if you've a half-decent Zephyr, Zodiac or convertible. You'll improve the car's

appearance, desirability and value by more than the numbers cost.

Buyers' guide

One thing that definitely hasn't changed over the years is the MkII's outer panel rusting characteristics although, in fairness, it must be stressed that they are more durable than many of their contemporaries (as solid as the pyramids compared to a PA Cresta!) and, in the case of saloons at least, there isn't often that much concealed frame corrosion; generally speaking the rot you see in the outer panels is the rot there is! New panels are, needless to say, virtually unobtainable and secondhand replacements are also scarce. Front wings are a particular shortage area. Bear this in mind before throwing away a tatty one; you may have no choice but to repair it. Different front wings are, of course, fitted to the four- and six-cylinder cars and all three models have different rear wings, although the arch section (usually one of the first places to rust) is sufficiently common for one repair section to cover all. In any case changing a complete rear wing is a very major undertaking; book time for just the basic job is about 20 hours.

Convertibles generally rust more than saloons as they are less watertight and also because the shells flex more. Check the entire body for rust, giving particiular attention to the sills, floorpan, sill/rear panel join and underbody chassis members. Have a look, too, at the power hood operating gear (if fitted). This uses brake fluid and a minor leak can lead to serious problems. It's also sensible to try a simple test for shell weakness. Open *both* doors and then close then one at a time. Do the test first with the car empty, then with one or two people sitting inside. If the doors foul the pillars when closed the car is probably weak (definitely if the car's empty) and repairs will be needed before long. With the car full, incidentally, it's normal for some sagging to be apparent and not a problem. Convertible doors and rear wings are different from saloons and well-nigh impossible to find.

Like their predecessors, the MkIIs had MacPherson strut front suspension. They are less susceptible than most later Fords to the usual inner wing problems but some will have been repaired as this example has. Not particularly pretty, but strong enough. Fitting new inner wings is a major undertaking (assuming you can get replacement panels in the first place) so make sure what you have is repairable; the metal's pretty thick by modern standards so welding new sections in should be straightforward.

Like many big saloons, MkIIs can be heavy on suspension; back end problems in particular are quite common and, because of the slightly 'tail down' rear styling, can be tricky for the novice to spot. As a guide, with the car on level ground and a nominal weight in the boot, the horizontal part of the wheelarch should be more or less level with where the rear tyre and rim meet. If the line's appreciably lower than this a rear end overhaul is needed. Although a MkII's ride is typically 1950s, softness should not be confused with tired shock absorbers and joints – and don't rely on the MoT test to pick up all defects. Box-based steering systems always have more play built in than rack and pinion set-ups but more than 2in play at the steering wheel is a sign of box wear. Individual parts are scarce but complete rebuilt boxes are available for from £100 although, if the worm needs replacing, the rebuild will be more expensive. Steering joints can also wear but most parts are available although track-control arm bushes can be tricky and complete idler arms are almost impossible to find although bush kits can still be bought. Suspension top mounting rubbers sometimes perish with age but remanufactured replacements are now available.

Scrapped MkIIs are a rare sight these days. Both these cars were too far gone to be restored when broken but that was some time ago and, today at least, one would probably be saved. Notice that the shells have been stripped of everything; it's all useful stuff!

Mike Staddon's Consul is immaculate under the bonnet. The oil-bath air cleaner was a feature of fifties cars; much cheaper than fitting a new paper element even if it is messier!

FORD MkII CONSUL, ZEPHYR & ZODIAC

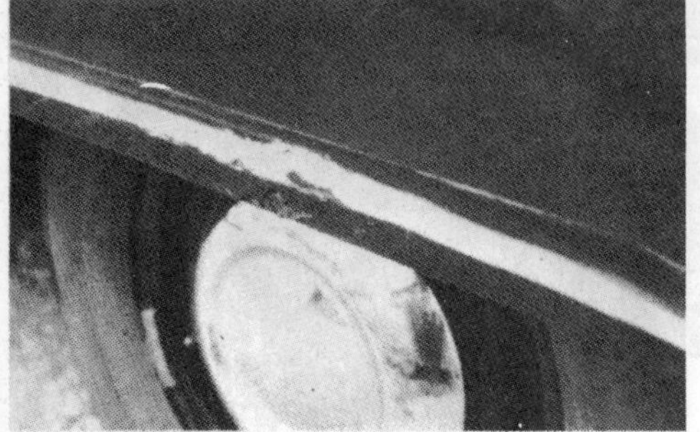

Wheelarch rust. To rectify this properly a quite large repair section is probably needed.

Both engines are reliable in the main and, given regular informed servicing, will have a very long service life. The most common telltale sign is fuming from the rocker box, usually at its worst after a long high-speed run or when the engine is pulling hard. The engines are perfectly conventional and DIY overhaul is straightforward. Some parts including oil pumps, rocker shafts and arms, cam followers and camshaft drive gears are hard to find; it's worth trying old-established remanufacturers though, as surprising numbers still have old stock tucked away and the machining should present no problems to a specialist. Five years ago the cheap way out of an engine blow-up was to buy a complete running but rusty MkIII Zephyr/Zodiac and swap the engine; that isn't so feasible now though as MkIIIs are also appreciating. Gearboxes are also generally reliable and failure usually follows a pattern; almost always the first sign of impending doom is jumping out of 2nd on the overrun. If this happens expect a gearbox rebuild within 10,000 miles. Some

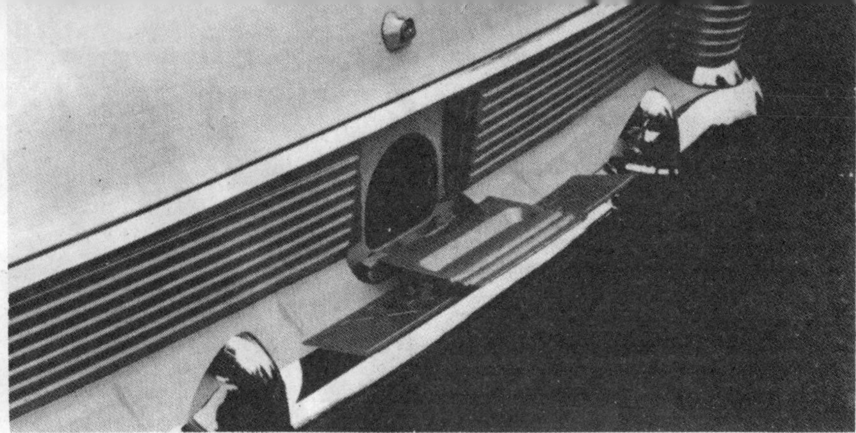

MkII Ford petrol filler caps were hidden behind the rear numberplate. Carefully-applied areas of trim distinguished the models; only Zodiacs had the bright rear panel.

individual parts, notably synchro rings, layshaft gear clusters and second gears are scarce.

Although, as one would expect, disc brakes are better than drums, the all-drum set-up, if in good condition, is certainly adaquate for the Consul's performance. However, drum-braked six-cylinder owners would do well to remember their systems' limitations! Most brake spares are readily available although front discs and servos are scarce; new calipers cost £120 each but it's often possible to have stainless inserts fitted and a piston refaced. Some Rover P4s used a very similar servo.

Bits and pieces

As will be obvious from the foregoing, there are still some parts shortages and, as the cars become more popular, so old stocks become used up. There has been a certain amount of remanufacturing though; items which we understand have 'come on stream' since November 1985 include door handles, window rubbers, hub caps, script badges and stainless steel exhausts and 'hockey-stick' manifolds. Then there is the Australian connection. MkII Zephyrs were extremely popular 'down under' and thanks to the climate many secondhand panels can be found in excellent condition. Ken Tingey of Ford 50 Spares (0202 679258) imports a container-load of panels every few months; as an example of price, first-class secondhand front wings cost around £150 each. Few purely Consul and no purely Zodiac and convertible parts come from Australia though; convertibles are even rarer and more sought-after there! Australian owners have also arranged remanufacture of some rubber items not yet remade here, including engine and gearbox mountings. □

Convertibles had three hood positions, lowered, raised and coupe de ville. The powered system (if fitted) raised the hood to the coupe de ville position as shown in this original Ford publicity shot; the front is then pulled across to the screen by hand.

Price Guide

Condition (see Price Guide)	1	2	3
Consul saloon	1800	1100	450
Consul convertible	6000	4000	1000
Zephyr saloon	3300	1800	475
Zephyr convertible	8000	4000	1000
Zodiac saloon	3600	2000	490
Zodiac convertible	8500	4000	1000

All these prices assume cars have non-suffix registrations, otherwise deduct 50% or £750, whichever is less. Estates should be priced as for saloons; despite their rarity they are still thought less desirable, in fact one dealer said that given a choice between buying a saloon and estate in the same condition at the same price he'd choose the saloon every time – so much for my prediction in 1985!